T5-AGO-966

Constructing Leisure

Also by Karl Spracklen

HEAVY METAL FUNDAMENTALISMS (*co-edited with Rosey Hill*)

SPORT AND CHALLENGES TO RACISM (*co-edited with Jonathan Long*)

THE MEANING AND PURPOSE OF LEISURE

Constructing Leisure

Historical and Philosophical Debates

Karl Spracklen
Leeds Metropolitan University, UK

GV
14
.S63
2011

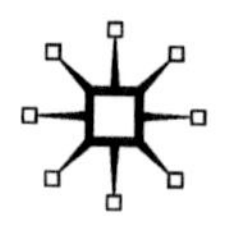

© Karl Spracklen 2011

All rights reserved. No reproduction, copy or transmission of this publication may be made without written permission.

No portion of this publication may be reproduced, copied or transmitted save with written permission or in accordance with the provisions of the Copyright, Designs and Patents Act 1988, or under the terms of any licence permitting limited copying issued by the Copyright Licensing Agency, Saffron House, 6–10 Kirby Street, London EC1N 8TS.

Any person who does any unauthorized act in relation to this publication may be liable to criminal prosecution and civil claims for damages.

The author has asserted his right to be identified as the author of this work in accordance with the Copyright, Designs and Patents Act 1988.

First published 2011 by
PALGRAVE MACMILLAN

Palgrave Macmillan in the UK is an imprint of Macmillan Publishers Limited, registered in England, company number 785998, of Houndmills, Basingstoke, Hampshire RG21 6XS.

Palgrave Macmillan in the US is a division of St Martin's Press LLC, 175 Fifth Avenue, New York, NY 10010.

Palgrave Macmillan is the global academic imprint of the above companies and has companies and representatives throughout the world.

Palgrave® and Macmillan® are registered trademarks in the United States, the United Kingdom, Europe and other countries

ISBN: 978–0–230–28051–9

This book is printed on paper suitable for recycling and made from fully managed and sustained forest sources. Logging, pulping and manufacturing processes are expected to conform to the environmental regulations of the country of origin.

A catalogue record for this book is available from the British Library.

A catalog record for this book is available from the Library of Congress.

10 9 8 7 6 5 4 3 2 1
20 19 18 17 16 15 14 13 12 11

Printed and bound in the United States of America

For Beverley (again)

University Libraries
Carnegie Mellon University
Pittsburgh, PA 15213-3890

Contents

Acknowledgements viii

1 Introduction 1
2 Philosophy of Leisure 14
3 Leisure and Human Nature 33
4 Leisure in Classical History 50
5 Leisure, Islam and Byzantium 67
6 Leisure in the Middle Ages 84
7 Leisure in Japan, China and India 102
8 Early Modern Leisure 121
9 Leisure in Modernity 140
10 Leisure in Historiography 157
11 Future Histories of Leisure 175
12 Conclusions 193

References 199

Index 223

Acknowledgements

Thanks to all at Palgrave Macmillan for their invaluable support, in particular Andrew James, Olivia Middleton and Philippa Grand.

1
Introduction

In the reign of the Emperor Justinian, the Late Roman Empire of the East had something of a rebirth (if not a renaissance in the sense of the word used by most historians today). In the capital city of the empire, Constantinople, Justinian built the *Hagia Sophia*, the church dedicated to the holy wisdom, which still stands above the Golden Horn in Istanbul. Justinian saw the church as a visible expression of his piety, and of his desire to unite the various theological factions of Eastern Christianity behind his own version of Orthodoxy. Along with the church and many other visible works of architecture, Justinian commissioned the compilation, editing and rationalization of the many strands of Roman law and created many important laws of his own (Watts, 2004). The Code of Justinian, like the *Hagia Sophia*, is one of Justinian's lasting monuments. More fleetingly, Justinian sent his ships into the Mediterranean and beyond to the lost Roman Empire of the West, where his soldiers retook the provinces of Africa, Italy and part of Spain. For a moment, with the defeat of the Italian Goths, it looked as if the Roman Empire was to be reunited and refreshed by the success of Justinian's generals.

Justinian, however, was never secure enough in his palace to be sure of a complete victory. His most famous and successful general, Belisarius, was lauded by the crowds of the capital city. But Justinian did not trust Belisarius. Through the campaign in North Africa, Justinian had been reluctant to give Belisarius reinforcements and fresh supplies. In Italy, again the emperor was slow to help his general, even when that general had taken Rome and was besieged behind its long walls. It was rumoured that Belisarius had been offered by the Goths to be crowned king of Italy. This rumour, according to the writer Procopius, had some basis in fact, but Belisarius had used the offer as a ruse to trick the Goths. Clearly, Justinian was afraid of Belisarius, and if the history of the Roman Empire was in Justinian's mind, it could be argued that he was right to be afraid. Since the year 68, when Galba succeeded Nero, the claim to the imperial titles was something that was seen as a reasonable career goal for ambitious generals.

But there was another reason why Justinian was afraid. The mood of the city was judged in the hippodrome, where the supporters of two rival chariot teams, the Blues and the Greens, were usually found in surly competition. Like many other Romans before them, the citizens of Constantinople were enthralled by the spectacle of the hippodrome, and the supporters of the rival factions knew that being a Blue or a Green meant so much more than mere cheering when the chariots raced around the circuit. Fathers enrolled their sons into the factions; wives were found within the right colour; money was won and lost through gambling; and even theology seemed to be connected to the Blues and Greens. Justinian had cultivated his favourites in the hippodrome, which in turn meant they were favoured across the city and the empire, in other cities where the factions divided people's leisure lives. His wife Theodora, if Procopius is true in his *Secret History*, had more reason than Justinian to favour one faction over another: she had been born into one, and then abandoned by that faction when her animal-keeper father died, to be enrolled in the other faction's entertainers.

So Justinian should have been secure, if the energy of the city's citizens, their agency, was constrained by the choice of Blue or Green. In fighting each other on the nights of races, the young men were forgetting something of the servitude they lived in; meanwhile, the other citizens who were in the factions could sit in taverns, curse their rivals and pray to any number of Saints to get their divine help at the next event in the hippodrome. It was normal for the emperor to be present at the hippodrome, where he could hear the complaints of his subjects and show his wise and Christian judgement. The emperor could reach the hippodrome direct from his palace, through a private passageway. This public forum allowed the emperor to control his subjects, to be mindful of keeping the balance between the Blues and the Greens, while at the same time giving favours to those who were useful to him and the empire.

That was the theory. In practice, the hippodrome's crowd could quite easily become a mob. And so it had proved to Justinian when he had detained members of the factions and allowed them to unite against him in a riot that turned into an insurrection. The organization of the two circus factions – ordinarily directed towards self-help, profit and winning – was combined in the *Nika* uprising against Justininian. From the hippodrome, the two united factions turned against the visible emblems of the emperor's reign, attacking his officials, then his palace. Rivals of Justinian were released from prison. The factions found leaders from families related to previous emperors. Fires started and spread across Constantinople. The supporters of the Blues and Greens – with their victory watchword for us symbolic of modern athletes and sneaker cultures – broke into the palace and pushed Justinian to a private harbour by the Bosporus. Procopius tells us that Justinian was ready to abandon the empire and flee into exile as a private citizen, but his wife refused to abandon her status as an Empress. She persuaded him to fight back – and with the help of Belisarius, who had appeared at his side, Justinian managed to hold on to his power. Belisarius and his troops entered the hippodrome and killed thousands of the Blues and the Greens.

That was why Justinian was afraid. He had seen the balance of the hippodrome overturned by the Blues and Greens fighting together. He had heard them shouting victory, seen them parading their would-be emperors through the streets. In public, of course, he showed no sign of weakness, only piety and grace. But in private, he must have shivered at the memory of the days of the riots, when he almost lost his empire – and probably his life – to the combined forced of the circus factions. Races continued in the hippodrome, and the Blues and Greens survived, albeit drained of resources and lives, but Justinian was evidently worried that a general like Belisarius might be someone whom the leisured layabouts of the hippodrome might choose to parade through the streets (Evans, 2000).

The circus factions of the Roman Empire are not new to critical studies of leisure and sport. For Crowther (1996) there is a danger in drawing too many comparisons between the circus factions and modern sports fans' violence. Drawing on the detailed account of the circus factions in Cameron (1976, 1979), Guttmann (1986, 1992) provides a detailed historiography of the conflict between the Blues and the Greens. His analysis seeks to demonstrate change in spectator violence: that change being the less violent product of the civilizing process (Elias, 1978, 1982; Elias and Dunning, 1986), which means hooligans do not massacre each other in the same way as the Roman fans did. Guttmann also plays with the continuity of violent spectatorship. In explaining, for instance, that 'in the Circus Maximus... the partisans of the Blues and the Greens... probably had their own sections, very much like the football fans of modern Britain' (Guttmann, 1992, p. 141), he draws a neat parallel between football hooliganism in the late twentieth century and that of the circus factions. In Coakley's student textbook *Sports in Society*, a general overview of classical sports history includes a brief discussion of games and sports in the empire that suggests (Coakley, 2003: p. 68):

> Chariot races were the most popular events during the spectacles…Spectators bet heavily on the races and, when they became bored or unruly, the emperors passed around free food to keep them from getting too hostile…This tactic pacified the crowds and allowed the emperors to use the spectator events as occasions to celebrate themselves and their positions of power.

Coakley's discussion is reminiscent of the Ridley Scott's egregious film *Gladiator* (2000), in which attention to accurate details of costume and background is countered by a typically Hollywood invention of history. In this artefact of popular culture, if not in actual second-century Rome, the Emperor Commodus is slain on the sand of the arena, jeered by the mob he had tried to placate with ever more extravagant gladiatorial displays. Coakley's Rome, like Scott's Rome, where leisure is seen as the diversion of the masses, is, of course, an old one. Juvenal, the Roman satirist, wrote that the people of Rome in his day were happy to be given dole (bread) and entertainment (circuses) as diversions from engaging in political debate. There is a weakness to this replication of Juvenal's view: Juvenal was writing in a particular historical moment, for a particular audience, for a particular purpose. If we blithely repeat his vision of Roman leisure as imperial hegemony, we face the danger of endorsing a narrow view of history. Guttmann is more sophisticated with his history and his sources, but his work is potentially problematic through its present-centred insistence on seeking the fairy story of figurationalism in the historical record. What Coakley and Guttman both do is fixate on a certain image of Roman games as sport, as some familiar combat watched by thousands of fans. It is a valid historical story to tell, but it is a narrow one. To understand the meaning and purpose of the circus factions, for instance, we need to look beyond simple stories of hegemony and civilizing processes, and examine wider issues of how and why societal and moral factors played a role in shaping the leisure of people in Rome and Constantinople, and how much agency and reason were involved in the construction of their leisure lives. We need also to examine the way in which those leisure lives were constrained by the social structures of the time, and the status of women, outsiders and slaves.

In other words, what is needed in any discussion of the circus factions and the games, or in any history of leisure, is an awareness of the philosophical problem of leisure: the paradox between freedom and constraint. In *The Meaning and Purpose of Leisure* (Spracklen, 2009, pp. 13–14), I outline this paradox of leisure at the end of modernity through an example drawn from teaching first-year Leisure Studies students:

> The arc of debate in a first-year class about the meaning of leisure is reflected in the growth and development of leisure theory and the discipline or field of Leisure Studies. Within the student discussion there are three ontologies of leisure: leisure as free choice in a world where leisure is defined by choice against other areas of life that are structured (for example, work); leisure as structurally-constrained choice (or no choice); and leisure as a completely free choice in a world where structures are breaking down altogether. These three ontologies are directly related to three epistemologies of critical studies of leisure, associated with the history of Leisure Studies as an intellectual study: liberal theories of leisure as freedom; structuralist theories of leisure as a (re)producer of social structures and unequal power relations; and postmodern theories of leisure (along with post-structural theories of postmodern leisure).

In that book, and earlier work (Spracklen, 2006, 2007, 2009) I argue that the paradox of leisure could be resolved by viewing leisure through a Habermasian critical lens. Jurgen Habermas speaks of two ways of thinking about the world, two rationalities, that in turn create human actions. The first way of thinking is communicative, which comes from human discourse, the application of reason, free will and democratic debate. Leisure, then, seems to be a human activity where communicative rationality is at work – we make rational choices about what we do in our free time. The second way of thinking is instrumental, which is what happens when human reason is swamped by rationalization, economic logic or other structural controls. Habermas' concerns with instrumentality are at the centre of his historical project (Habermas, 1984[1981], 1987[1981]), which maps the rise of communicative rationality and its lifeworld of human discourse, and the struggle to keep communicative reason afloat under the stress of the instrumentality of late modernity. So leisure, from being something communicative, follows the inexorable logic of capitalism to become something instrumental. In the concluding paragraph of *The Meaning and Purpose of Leisure* (Spracklen, 2009, p. 159), I argue that:

> Critical studies of leisure, following the pessimism of Adorno and Gramsci, can and indeed should be maintained as a means of identifying and understanding the increasing dominance of instrumental rationality in society and culture. Leisure as a meaningful, theoretical, framing concept; and critical studies of leisure are a worthwhile intellectual and pedagogical activity... Indeed, leisure is the part of our lives where the tension between freedom and constraint – agency and structure, resistance and control – is most visible, so understanding leisure is even more essential as the world and its societies become increasingly commodified and ordered. Following Habermas, examining leisure actions can help us understand the conflicting pressures of instrumental control and individual will – and in doing this, critical studies of leisure can and should continue to play a central role in understanding society.

This book is intended to be a sequel to *The Meaning and Philosophy of Leisure*, though *Constructing Leisure: Historical and Philosophical Debates* can be read without ever reading my first book on leisure. Where that first book is focussed on leisure at the end of modernity, and the effect of globalization and postmodernity on leisure, this book looks back at the meaning and purpose of leisure in the past. It is a history and philosophy of leisure. But this is not a simple social history of a leisure form, such as Borsay's (2005) book on British leisure and class identity. It is not enough to write a history of leisure on its own – in fact, it is impossible without engaging in the debate about what counts as leisure (in the present and in the past). It is necessary to examine leisure and theories of leisure in historiography, critically, and through the lens of philosophy. This book's aims, then, are twofold: firstly, to engage with academic debates about leisure in history and philosophy, which will lead to a strong critique of the narrow focus of previous historiography and social theory; and secondly, to provide a much broader chronological and geographical scope for problematizing leisure, which allows for both a more balanced analysis of the meaning and purpose of leisure, and a comparative exposition of that meaning and purpose in context (Roberts, 2011).

This realization that you cannot do history without philosophy is accepted in academic discourses around the meaning and purpose of science. The history of science and the philosophy of science, though clearly delineated scientific disciplines with their own corpus of knowledge, are best understood as refracted elements of a meta-discourse of meaning exemplified by the sociology of knowledge, which ties the two together: the history and philosophy of science (Golinski, 1998). For example, in 1543 Copernicus, in his defence of the heliocentric universe described in *De Revolutionibus*, wrote that 'all this is suggested by the systematic procession of events and the harmony of the whole universe, if only we face the facts, as they say, with both eyes open' (Copernicus (1947 [1543]), Book One, Chapter Nine, cited in Kuhn, 1957, p. 154). According to Copernicus, there was a physical, heliocentric system, which he put forward as a true account of the universe to replace the Aristotelian, geocentric universe. However, the mathematical system adopted by Copernicus to account for the phenomena of the movement of stars was similar to the one used by other astronomers created by Ptolemy in the late Hellenistic era. This system was used by astronomers to make predictions, but it held no metaphysical or physical reality. For the philosophers, the Aristotelian model was the 'true' model – the astronomers merely used the complex Ptolemaic model because of its usefulness in making predictions (Rose, 1975). So it comes as no surprise to see Osiander writing in the preface to Copernicus' work that the heliocentric model was not physically true, but mathematically useful (Kuhn, 1957, p. 187). Duhem has argued that Osiander was right to make this instrumentalist intervention, that there was no justification for anyone such as Galileo to make the claim that the Copernican system was really real – in Duhem's account, Galileo should have listened to Cardinal Bellarmine, who urged such an instrumentalist approach and saved himself a lot of trouble (Duhem, 1969). Constructivists argue that we should look to external influences and interests, as well as practice, to write meaningful narratives in the history of science (e.g., Barnes, 1977). Copernicus was strongly influenced by the philosophy of scholasticism, hence his commitment to perfect spheres (Debus, 1978). But he was also influenced by the rediscovery of Neo-Platonic texts in the 100 years leading up to 1543 (as witnessed by his commitment to an underlying physical reality and his Neo-Platonic reverence of the sun), a rediscovery effected by the fall of Constantinople in 1453, the growth of mathematical disciplines in independent academies and the invention of the printing press (Eisenstein, 1983; Cohen, 1994). The Reformation and Counter-Reformation played a massive role in the transmission and initial suspicion of heliocentricity. Galileo could not adopt a pragmatic approach to heliocentricity that reconciled philosophy with theology because he believed in a physical reality verified by observations he had made through a telescope. To understand what was happening up to and after Copernicus, then, one needs to understand the philosophical debate about what constituted good natural philosophy, the debate about what was real and how that reality could be discovered and justified, and how that debate was situated in wider discourses about knowledge, power and God.

Writing a history of leisure, then, entails writing a philosophy of leisure – and any history needs to be a philosophical history as well. That is the purpose of this book. It will provide an account of leisure through historical time, how leisure was constructed and understood by historical actors, how communicative reason and free will interacted with instrumentality at different times, how historians have reconstructed past leisure through historiography, and finally, how writers have perceived the meaning and purpose of leisure in alternative histories. This is, then, a more considered and *philosophical* history of leisure than that sketched out by Blackshaw (2010) in his polemic for the *uniqueness* of 'liquid' leisure in the twenty-first century; or the ahistorical musings on leisure and Eliasian civilizing processes found in Rojek (2010).

What is history? If we are to address historical issues within our particular field of research there has to be an understanding that, in many cases, history is indeed bunk. Or rather, there is a danger that history can be bunk if we charge blindly into the past seeking events and evidence that we find relevant. The problem is not the past *per se*, but our use of history, and our writing of it. Far too often, history is presented as a simple chain of causal events in the past, a preparation for the present that is self-evident to the researcher (Zammito, 2005). Yet, this is blatantly not the case. If we take the past as a creation of the people and culture of that relevant time, we are given a far more complex view of a past made for their time and not ours. And, if we are to take a postmodern view, the best that we as historians living in the present can do is create histories amongst ourselves, and the idea of a History (with a capital letter to signify the Grand Narrative) that is accessible becomes a lost cause. Too many events happened in the past – all we can hope to do is look to discourses situated in small parts of time and space: any judgement on the past (that is, comments that 'this was good, that was bad') by historians living in the present is an anachronism that must be avoided.

This book does not claim to be the history of leisure, or even the philosophical history of leisure. Such narratives are outside the scope of any research, as well as open to anachronistic criticism. Indeed, the very term 'leisure' is problematic if used too casually in history, as its meaning and purpose changes from the early modern period onwards (Bramham, 2006; Spracklen, 2009). What this book will approach are the areas of history that are directly relevant to understanding social and philosophical problems about leisure's meaning and purpose, as an aid to my understanding of the field of leisure studies. At all times in this historical analysis, I will attempt to utilize an awareness of the historian's role in history making, by providing alternative explanatory frameworks to what was 'actually' happening. That said, I will also provide my own interpretations of the relevant historical events, and show how they are pertinent to my own research agenda: that of providing a critical synthesis of the communicative meaning and purpose of leisure for human societies.

However, before I look at these two aspects of my historical discourse, I have to be sure of the problems that arise with all discourses that discuss the past. In the next section of this introductory chapter, I will discuss the dangers of Whig history and the attendant problems arising from our present-centredness and the postmodern view of history being a collection of opposing discourses over what counts as 'historical truth'. I will then suggest what has to be done in any historical discourse to avoid both conservative ideologies deciding what counts as history and, for example, thinking William the Conqueror invading England in 1066 is all you need to know about anything.

Whig history: present-centred, present-directed, progressive

The first pitfall in history writing is that the past is vast. If we make a simplistic definition that goes something like 'The subject of history is all things that are located in the past, being the time before the contemporary circumstances', we are left with a huge amount of subject matter. There is no feasible means by which anyone could hope to know everything about the past. Realistically, then, history is approached by adopting themes, subjects or interests, and exploring them in the past. In my case, my interest with the past has already been stated. I have a specific interest in one small part of history, a history of leisure, done to explore the philosophy of leisure: specifically, the concept of leisure as a place of tension between Habermasian communicative rationality and instrumentality (Habermas, 1984[1981]; 1987[1981]).

One must suffer the consequences of any historical departure in research. There are problems highlighted by historiography that will now be discussed. Then I will proceed to raise the problem of the relativity of knowledge and how this creates problems for our understanding of history. These issues are the centres of intense debates in their respective disciplines, and their relevance in this paper may be questioned. However, it is vital that we engage in these debates, or at least acknowledge them, as they pose questions over what is good and what is not (close to actual truth) historical discourse.

'Whig history' is a term first coined by Herbert Butterfield. He defined it as the tendency of historians to see the past as the story of conflict between progressives and reactionaries, in which the progressives win and bring about the modern world. He suggested that this was to overestimate the likenesses between present and past and to ensure that we always intend the consequences of our actions (Butterfield, 1968[1931]). Whig history, then, supports a progressivist view of history, as if the entire past was a build up to the present, a supporting act to the events of our time. Although the term may be thought of as ethnocentric, since Butterfield was only interested in history as Western history, the ideas behind it remain powerful. Whig history becomes equated with the modernist paradigm, and all historical discourses that use the past to prove a point in the present. Whig history is the propagandist history of ideologies; it is the use of history to justify the actions and interests of the Whig historian. In the case of the Split in 1895 of the Northern Rugby Football Union from the 'official' union in Twickenham, for example, Whig history has been perpetuated by both rugby union and rugby league historians, who only see what they think justifies and supports their view of what sport (and their sport) is about (e.g., see Moorhouse, 1989; Macrory, 1991; Collins, 1999, 2006). The historian begins to interpret the past by his or her contemporary interests (Skinner, 1969). Also, the Whig historian, in writing this particular history, judges the life of the past so that, for instance, Newton becomes a father of modern science (Westfall, 1977), and all the natural philosophy that comes before is seen as irrational or stupid, which plainly was not the case (see, for example, Pumfrey et al., 1991). Whig history is essentially progressive, present-directed and present-centrist, in that it considers the past through the values of the present (so that what the Whig historian decides is important in the past comes from what he [it is usually androcentric] feels is important in the present). Joyce (1995) questions the use of class in social history as a present-centrist term that does not do justice to the different discourses of the past. Whilst this is a rather brave attempt at directing the attention of historians away from a dangerously Whiggish over-emphasis on class (a present-centrist concept), the importance of class is undeniably important both in the popular imagination of the late nineteenth century, and in the analysis of society from then to the present. That said, it must be realized that class is not the sole definer of social relations – Joyce's concept of 'the people' in the democratic process is another delineating concept in social relations, and in relation to my work masculinity and gender work in opening up history (e.g., Hargreaves, 1994).

It is clear that Whig history, which is similar to the Grand Histories of the nineteenth and early twentieth centuries (see Wells, 1936), provides a distorted view of the past. It provides a view of the past as seen through the desires and interests of the historian. Such history naturally becomes a thing to be avoided, and most historians today are sensitive to the issue. Yet, even with a sensitive awareness of the problems of Whiggism, we are still perpetrators of distortions, since we are writing our historical discourses in the present. Our historical discourses are inevitably present-centred (Ashplant and Wilson, 1988; Wilson and Ashplant, 1988). In addition, our interpretive skills and our language are also centred on the present (Skinner, 1969), as well as our use of the discourse. In a sense, then, writing true history becomes impossible, as our inherent present-centredness will always get in the way of any objectivity.

The linguistic turn and the problem of relativism add more pessimism to the debate. For if our desire is to look at the history of our particular subject, we are told it is a hopeless task. Not only is the past seen as another symbol (or symbols), with a myriad of different interpretations (Baudrillard, 1988), but relativism states that we cannot know which historical discourse out of the different discourses provided is the right one. In thinking about the history of rugby league again, do we believe rugby league sourced histories, rugby union sourced histories, figurationalist-sourced histories, Marxist-sourced histories or something else? Postmodern historiographers suggest this is a useless task, as there may not be a truth anyway, merely a parade of localized historical discourses (Foucault, 1970, 1972).

We have two options: abandon history, or accept the realness of the past and hence provide a reference for all historical discourse, so that it then becomes possible to say whether discourses are good or bad (in relation to this 'real' past, via credible discourses made more consistent through a rejection of distorting ideologies such as progressivism). Whilst I tend to a relativistic view of knowledge, the fact I am engaging in historical discourse suggests I am following the latter course. But I am not merely dismissing other discourses to present mine as the truth. Rather, I am looking at the secondary sources and finding out what they say. The way around the problem of present-centredness is to look at discourse production (Lorenz, 1994; Bentley, 2006). Who has written the history? Why? What were the sources? Does this agree or disagree with other discourses on the same subject (Ashplant and Wilson, 1988; Wilson and Ashplant, 1988)? In response to the question of what the past really is, or whether we can ever get true access to it, one can only make a leap of logical faith. If, as we have seen, accounts of history that are Whiggish (or ethnocentric, or androcentric or otherwise progressivist) are bad history, then the removal of all these habits from our historical discourses will make them better, that is, more faithful to the events of the past. If we then accept that we are interested in specific parts of history, that we have a purpose which is overtly acknowledged, and that to 'know true history' is impossible and mis-directional, our discourse will be far more credible. I cannot help but write my historical discourse with my interests in mind – why else would we choose to write history? But I can be aware of that and the danger of excessive truth-claims. Hence, I am looking at the historical discourse of the people who wrote the primary and secondary sources, and I am watching out for the pitfalls that make good history bad (Bentley, 2006).

My choices for this book, then, are inevitably personal and partial. There needs to be a summary of discussions about leisure from philosophy, and some discussion of the nature of the history of philosophy. This is the content of the next chapter. The rest of the book provides a philosophical analysis of the history of leisure in a number of specific contexts. The historical discussions need to be useful in the exposition of the meaning and purpose of leisure – to allow me to write this book. There needs to be a rational sifting of chronology, culture and geography that allows the book to be pertinent and relevant to its readers. The history cannot be exhaustive. Inevitably, there is a need to balance ethnocentric, Western sources and the limitations of what is available from the pre-history, archaeology and un-translated sources of non-Western cultures. I believe that I have achieved this careful balancing act of providing useful, relevant historical contexts for leisure in which arguments about the philosophy of leisure can be made. My historiography, then, begins in Chapter 3 with an exploration of leisure in pre-historical times and pre-historical cultures. Chapter 4 concentrates on the Mediterranean world of Classical Greece and Rome, where there is a wide range of primary and secondary sources to guide us. Chapter 5 follows the Roman Empire into the Byzantine and Islamic cultures that succeeded it. In Chapter 6, the chronological order is continued and leisure in the Middle Ages is explored, drawing on the relationship between Christendom and Islam and arguing for the influence of the latter on the former's courtly leisure culture. Chapter 7 will examine leisure in the history of India, Japan and China in pre-modernity. Chapter 8 will return to

Europe and focus on the Early Modern period. Chapter 9 will then take the chronology into Enlightenment, the industrial age and modernity. Chapter 10 will explore the theme of leisure in writers from the Early Modern period onwards, engaging in particular with modern historiography of leisure in the work of writers such as Dewey, Marx, Weber and Veblen. Finally, Chapter 11 will examine modern history, postmodern history and alternative histories. A concluding chapter will then summarize the endeavour, return to the histories of leisure discussed in Chapter 10 and identify future research projects within the framework of philosophical history, and historical philosophy, of leisure. In this synthesis, much, of course, will be missed. The past is full of things that happen. But in making this historiographical selection, it becomes possible to write a history and philosophy of leisure.

2
Philosophy of Leisure

This chapter provides a summary of discussions about leisure from philosophy, drawing on classical authors such as Plato, Aristotle and the Epicureans and Stoics, as well as more contemporary work from Huizinga (2003[1944]), Morgan (1976), Suits (2005) and Holowchak (2007), and some discussion of the nature of the history of philosophy. The first section of this chapter begins with the above-mentioned classical authors, and their ontological, epistemological and moral accounts of the meaning and purpose of leisure. I will argue that although there is a commonality in those accounts about leisure as the preserve of the learned, wealthy and free man, the four different philosophical schools established different meanings and constructions of leisure. Following Plato, for example, true leisure (*epistemic* leisure) becomes a metaphysical Form, of which our *doxic* leisure activities inevitably become a mere reflection: hence, it becomes possible to distinguish hierarchies of good and bad leisure, with rational discourse and ritual sports becoming the most noble. For the Epicureans, anything that helped fulfil the need for happiness and satiation counted as good leisure: so for them, the best leisure forms become those which give some kind of satisfaction, whether that is the pleasure of climbing a mountain, having sex with a lover, or finishing a bottle of wine. The second section of this chapter will sketch a history of philosophy from Hellenic times through the Scholasticism of the Middle Ages, Cartesian dualism, Kant's attempts to build a moral framework of everything, Hegel and History, and the professionalization of the discipline in the twentieth century. In this chapter, I will consider the accounts of leisure given by key philosophers, from Aquinas to Wittgenstein, and the context of their philosophy. I will end this chapter with a section discussing modern philosophical accounts of leisure, in particular, the work of Huizinga on the essential playful nature of humanity, and the work of Suits on the ideal nature of sport. Before that, however, it is necessary to understand some of the key issues of ontology and epistemology that have shaped philosophical inquiry over the years.

Some issues that trouble philosophers' dreams

Here are a few issues that pose dilemmas in research and any inquiry into the truth and the nature of reality. These problems have to be addressed – ignore them and you go against the spirit of philosophical inquiry that has led us all here in the first place (even though ignoring them is the best way to progress within the academic factory of patronage and traditions).

Ourselves and the mind/body problem. We are the starting point of any inquiry, as well as the end point. But what is *our* nature? I have, I believe, direct access to my consciousness. I have no access at all to yours. Is my mind reliable? Is it independent of the reality I am investigating, or does the mind produce this reality? What is my mind? Where does my belief in my own consciousness come from?

God and/or determination. Is everything, or a part of the sum of existence, predestined by some higher authority, whether it is a supreme being (or beings) or something similar to the will to power? Or are we just a random accident, a possibility that was actualized from an infinite range of possibilities? How do we account for any predetermination, or its absence, in theorizing and conceptualizing our ontological understanding?

Reality. What do we mean by reality? Whose reality? Does a reality exist independent of the workings of collective human minds? If so, can we as humans see this reality and understand it? Or are we limited by our own reasoning, and hence producing ontologies of our own that either have no relation to the really real, or are pale imitations of the really real? And if this is so, how do we determine which 'worldview' of reality is the best? Or should we even forget about this really real reality that may or not be there and which may or not affect the creation of paradigms of shared assumptions on the nature of an *internal* reality?

Bias. We, as researchers, claim to be revealing something about reality, about the real world. Are we really doing what we claim, or are we producing ideas that stand up only in one epistemological framework, the one of the western academic world in which we operate? And how can we define this framework? How are we enculturated into it? How do new ideas and paradigms gain credibility inside this system? How do claims go from false to true?

Epistemology. Many theories seem to explain the same data. How are empirical results related to theorizing, if we forget for a moment that we have already questioned the place these results supposedly come from? How are theories proven as true? Why, historically, have valid ideas later become invalid, and invalid ones valid? Who decides what is added to the commonwealth of knowledge? There is a problem with the absolute truth, the logical form to which knowledge aspires. Not only does the sociology of knowledge show truth claims to be made on the basis of non-logical rationales, and that scientific truth does not equate with really real truth (suggesting that the truth claim is bound to *a priori* assumptions), but it also suggests that *really real truth*, as first explored in Socratic philosophy, lacks a rigorous defence. In other words, Socratic truth is really valid only because its defence believes that Socratic truth is really valid. Logic, therefore, seems to become as dependable on faith as reality and religion, which is not what philosophers or historians want when they make truth claims.

Relativism. There have been a plethora of sound rhetorical arguments against the idea of one truth, one reality. Relativism and postmodern philosophy speak of truths dependent on cultural specificity, of discourses, that are incommensurable. That is, one claim cannot be judged truer, or more real, than another. This is an approach sensitive to all the questions put forward in this argument, as well as an approach that realizes that different people do have different worldviews, that are influenced by politics, language, cultural background and other factors. Most of the arguments in this ontological discussion support a relativistic approach to reality and truth. However, this is the crux of the dilemma. If relativism says all arguments have equal validity, why are the arguments of relativism truer in the eyes of relativists than orthodox scientific arguments? In other words, to support a relativistic worldview, one tacitly accepts that relativism is just as tenuous in its truth value as the worldviews it opposes. Or worse, since you can work with any worldview depending on your interest and culture, what is wrong with being rational, Socratic, Western and scientific? Clearly, these faults of relativism have to be addressed; otherwise, relativists will be continuously hoisted by their own petard.

Priority. In philosophy, it is clear that ontological things (reality) must come before epistemological things (knowledge, how we get it from reality). However, this is not how we seem to work. We have already seen how, in this discussion, epistemology has come before (and helped create) our ontological understanding. This is also true of all knowledge production. We have to learn to learn before we learn about 'reality'.

Why have humans got the two the wrong way round? Why do we cloud our judgement before looking at what is real and true? How do our *a priori* assumptions shape a belief in an *a priori* reality? Does our enculturation, our epistemology, create our ontology?

Language. We are bound by language in describing and conceptualizing. How is the problem of a limited language overcome? Or does our language decide what is in our worldview? How can we try to understand an external, meta-reality, whilst operating in the symbols and meanings of a particular paradigmatic dialect? What do we learn by way of meaning when we learn words? Is meaning in usage or definition? How can we tell, and more importantly, how can we differentiate between meanings from usage or meanings from definition? How does this affect our own inquiries? How do I know that what I mean to say will be understood in the same way by you? And if there is a difference of understanding, what is the *proper* understanding of the words? For example, what do I mean, what do you mean and what do we understand, when we talk of reality?

The nature of any standard academic argument is to try and convince the reader of the credibility of said argument. To do this, the writer uses a rhetorical style of appealing to precedence, making recourse to the elders who came before who have already proved their theoretical validity. For a novice to write a paper without any appeal to these accepted fonts of knowledge is seen as 'bad' academic writing, even though we are supposedly in the business of thinking for ourselves. The concept of theory, then, is extremely simple. Much has been written about what we mean by theory, but I will skirt past those arguments to try and situate a meaning as it is understood in the commonwealth of linguistic interpretation. A theory is an idea, one that is related to a particular set of data, or part of reality (let us call it a 'field'), that claims to shed some light on what is happening in that field. There is a debate over whether a theory is a description or an explanation. The former is a theory that merely provides an account of the field, whereas the latter goes some way to interpreting beyond what has been seen. The latter may also serve as a predictive tool, using the explanation of the field to try and predict what may happen elsewhere. Yet, one can immediately see a problem with these fine distinctions. Whether a theory is descriptive or explanatory is usually the decision of a theoretical critic. Of course, the one he or she favours will be the powerful theory, the one which explains in analytical language the critic understands, and which supports a larger theoretical framework which serves as a predictive tool. But any theory has to be explanatory. By describing something, we are trying to explain what is going on. Our language is important. In a descriptive piece, we may not use the analytical language that identifies a particular epistemological framework, so naturally the critic will scorn it. Even so, we are using some kind of language. We are using some kind of analytical device. I have already suggested that everything we do, including this chapter, is only real through the language of our particular epistemological framework. In descriptive theory, we are still bound by the theory-ladenness of our language. What is being described is a reality, not the reality, of a situation – in other words, what is at stake is the epistemology, not the ontology. The dichotomy of description and explanation is a tool used by the academic world to bracket out what the critic sees as bad (that is, theory that does not support a particular bugbear) from good.

Aristotle dealt with many aspects of knowledge and knowledge production, but at the heart of his entire project (and it was a conscious project aimed at finding out everything – Aristotle was not known for his modesty) were the concepts of logic and Truth, theoretical concepts that had their own internal validity and reality. Truth, the nature of a thing or concept which could be described as true by all, was a hard thing to come to. He recognized that people believed in many different and opposing things, and he wanted to bring consistency to his project in the manner of many other Greek philosophers. In this way, Truth could be distinguished from Belief. The way to Truth was rational and rigorously closed – it was deductive logic. Although logic can be shown up for its external inconsistencies, there is in deductive logic an internal path that cannot be denied: *if* all men are mortal, and *if* Socrates is a man, we can deduce that Socrates, given the above, is mortal. There are no two ways about it. The third statement is *True*, according to the internal logic (the content of the syllogism is not, in this essay, important). Aristotle successfully describes a concept of Truth, as well as a type of logic, that form the basis of all the arguments that rage hundreds of years later. The essential problem which his legacy gave was the uselessness of deductive logic, and the tantalizing glimpse of a way to the truth. The uselessness of deductive logic does not stem from the mechanics of the argument, which are sound, but the tight and impossible conditions that have to be available before deductive logic can be used. Human beings seem instead to work with inductive logic – if we did follow a deductive logic then we could all breathe easy. Yet, the glimpse of truth was there, though it seems further away than ever. This is why science claims to follow a deductive method, and why Popper (1961) suggested that science works through deductive falsifying (though the mechanics of Popperianism are fatally inductive, too – see Maxwell, 1972), and why we have a crisis of failure, and why some people deny truth at all. Or, to return to charted waters, why relativists make their claims that since nothing can be proved true, and there is no way of comparing challengers to the Truth, we may as well work with a relativistic view where each Truth becomes just one truth next to another in a multitude of discourses. The problem inherent in the project of knowledge production is the problem of induction: we cannot have all the facts to help us decide the truth. To say the sun will rise tomorrow is not True, in an Aristotelian sense. We cannot know for sure what will happen in the future, nor can we be sure of the regularity of the Universe solely on the basis of our observations so far, whether our observations are from experience or from some theoretical framework like Marxism. All we can do, then, is to try to explore the history and philosophy of any topic to begin to understand the meaning of particular concepts, the discourses and symbols evoked and challenged, the controversies and conversations – what Foucault (1970, 1972) describes as the genealogies of meaning. For this endeavour, then, we have to begin with the discourses about the meaning and purpose of leisure.

Philosophy of leisure in ancient Greece and Rome

The meaning of leisure clearly is something to do with ontology, with metaphysics – we ask the questions: what is sport, or what is leisure? This becomes purpose when we ask what leisure or sport is for. Purpose is the ethical or moral function or role of something. The two are related – but often confused. Bentham, for example, in *An Introduction to the Principles of Morals and Legislation* (1996[1789]), tried to understand the rules of society in a rational, Enlightened way, developing his famous utilitarian theory of ethics. This utilitarianism is a philosophy of statistics and measurement. But Bentham was also a liberal with a strong sense of good (reading) and bad (drinking) in society: hence, utilitarianism's consequentiality is posited on an assumption that there are good and bad choices, good and bad actions, that can be measured empirically. That measurement is related to purpose – what is the action for? Who benefits from the action? Is there a net gain for society? We could ask – what is leisure for? Who benefits from leisure? Is there a net gain for society (from such leisure)?

The Ancient Greeks were the first philosophers, and the first to write down their thoughts on leisure and everything else (or, at least, the first whose writings have survived). Different Greek cities had different ways of defining belonging, making decisions, sharing power and educating young people, which included traditions about leisure and sport. In Sparta, for instance, the male children of the elite were taken away from their families and raised in a communal way as warriors. Their development involved a strict regime of sports and exercises, culminating in martial training. In other cities such as Athens, the male children of the elite were instructed by sophists and philosophers in the arts of thinking and learning (Braund, 1994; Rhodes, 2003; Fox, 2005; Blackshaw, 2010). The right way to live, and the best way to organize society to enable a city to prosper, was crucial for Plato. In *The Republic*, his tutor, Socrates, debates the culture and structure of a utopian republic, ruled by kings trained in philosophy (Plato, 2007). Plato uses *The Republic* to articulate his idea of the Ideal Form: this work is not just an attempt to articulate a perfect city-state; it is a representation of an Ideal Republic, the perfect Form of which could only ever be partially reflected in the day-to-day realities of our imperfect world by our imperfect senses. Plato's Ideal Republic has a measure of gender equality because of the cosmological story of the Divine Unity, out of which genders were split (Kochin, 2002). And the Ideal Republic has clearly defined leisure activities for different classes: education and training for the philosopher-kings; formal versions of instrumentalized leisure such as games for the other free men; and some free time from work for the lower orders (Balme, 1984; Braund, 1994; Young, 2005). Leisure in the Ideal Republic reflects Plato's own philosophy of leisure – the moral goodness of 'proper' elite leisure such as sports and intellectual inquiry, against the messy and immoral world of the shadows (Young, 2005).

Aristotle's philosophy broke with his tutor Plato. He believed in the good life, but rather than argue for something that existed in an Ideal Form, he suggested that the good life could be found through careful observation and argument – this commitment to inductive reasoning was found in all his writing that has survived. Naturally, he believed that his own life was a fair reflection of the best sort of life (Owens, 1981; Balme, 1984). His view of good leisure and sport was similar to Plato's: ritual sports and games, music and plays, genial debate in the town square and listening to the recitation of books (Balme, 1984). But he came to that view of leisure from a comparison of Athenian cultural and political life with other societies. Aristotle believed in the sanctity and naturalness of local traditions and leisure practices, in the value of leisure as a means of binding together communities, but also in the freedom of elite men to choose the particular leisure forms they practised (Owens, 1981).

The Epicureans followed the philosophical works of Epicurus (O'Keefe, 2009), who argued that the good life could be enjoyed through the cultivation of taste, both cultural and intellectual. The stereotype of the Epicureans as gourmands, gorging themselves on wine and food, is an old one, and there was some truth to the stereotype. Some Epicureans did believe that taste was something to be cultivated literally (Jones, 1979). However, most Epicureans believed in applying reason to exploring the boundaries of cultural and intellectual life, and establishing the proper way to enjoy all things in life. This meant, for example, understanding the balance of harmonies in music to better enjoy the sound; or establishing the proper way to train a body of a man to be strong enough to win at wrestling (Jones, 1979; Warren, 2006, 2009). While not as influential as the Platonists or the Aristotelians, the Epicureans influenced educated Greek and Roman pagans to see in their everyday leisure and cultural habits a refinement of taste and civilization (O'Keefe, 2009).

The Stoics were the most popular philosophical school in Hellenistic and Roman times (Strange, 2004; Irvine, 2008). The Stoics believed in a good life of moderation and contemplation, combined with detached reason. They were a strong influence on the Roman ruling classes. The Emperor Marcus Aurelius, in his *Meditations* (2006), espouses Stoic virtues of restraint, moderation, tolerance and wisdom. These were virtues said to be the mark of good living and good men. Marcus Aurelius's correspondence with his old Stoic tutor Fronto has survived and been passed down through the ages as an example of the application of Stoic philosophy to the practicalities of everyday life, politics and culture. In these letters, and in his *Meditations*, Marcus Aurelius discusses moderation in leisure practices: the inanity of dice; the folly of gambling; and the dangers of excess drinking and sex (Birley, 1993). There is also a sense of weariness about the plebeian spectacle of the games – as a politician, Marcus Aurelius recognizes the hegemonic power of such instrumentalized leisure, but as a philosopher and aesthete he is tiresome of blood-lust and unreason. For Marcus Aurelius, leisure time is best spent contemplating his place in the world, reading good works of philosophy and history, and preparing his mind and body for the labours of running the Empire.

Although there is a commonality in those accounts about leisure as the preserve of the learned, wealthy and free man, the four different philosophical schools established different meanings and constructions of leisure. Following Plato, for example, true leisure (*epistemic* leisure) becomes a metaphysical Form, of which our *doxic* leisure activities inevitably become a mere reflection: hence, it becomes possible to distinguish hierarchies of good and bad leisure, with rational discourse and ritual sports becoming the most noble. For the Epicureans, anything that helped fulfil the need for happiness and satiation counted as good leisure: so, for them, the best leisure becomes that which gives some kind of satisfaction, whether that is the pleasure of climbing a mountain, having sex with a lover or finishing a bottle of wine.

A history of philosophy of leisure

The second section of this chapter sketches a history of philosophy. The schools of philosophy of the Classical Age did not survive – at least in Western Europe – the rise of Christianity and the fall of the Roman Empire. Christian theologians took a dim view of the uncontrolled and sinful leisure pursuits of the pagan past, and distanced Christendom from amoral (and immoral) leisure activities (Chadwick, 1993; McDonald, 1998). Saint Augustine (2002), for instance, argued that pagan leisure activities were the work of the devil, and recommended to Christians a life of prayer and contemplation. In the Byzantine Empire, the Emperor Justinian's codification of Roman law diminished sexual and other leisured freedoms (Evans, 2000). Philosophy survived through the work of monks and collectors such as Cassiodorus, who gathered and copied manuscripts from around the West (Jones, 1945). In the East, the closing of the pagan schools of philosophy saw the drift of ideas, individuals and their work into Persia, where the books became a part of the Islamic world's inheritance. In all these places, a tiny portion of the work of Aristotle survived as the exemplar of Greek philosophy: Muslim philosophers commentated on this work, but there was no systematic development of philosophy as a discipline until the re-emergence of Aristotelianism in the Scholasticism of the Middle Ages (Hannam, 2010).

Western theologians and philosophers saw in Aristotle a way of accounting for the physical and divine, and the natural relationships and hierarchies of the world (Grant, 1997). Thomas Aquinas (2003) famously used Aristotle to provide a Catholic theology of God's omnipotence and humankind's free will. God is Aristotle's First Cause, the Prime Mover of popular discourse. The hierarchy of feudal society is predicated on a divine tolerance of inequality: just as dogs are closer to God than the fleas that bite them, so humans are higher than dogs, and priests and lords higher than merchants, and men higher than women, and farm labourers higher than actors and minstrels. Thomism legitimized the belief that every person had their place in a divine order (Southern, 1990), but it also justified individual salvation through grace. Every soul inhabited a human body and animated it with free will – our path to heaven was ours to make through following the rules of the Church and the commandments of the Bible. Aristotle's natural philosophy, then, became a dominant theme in the medieval philosophy of Scholasticism (Grant, 1997; Hannam, 2010) – and brought with it a view of free will that gave individuals the agency to choose the right (Christian) leisure activities and be saved, or the wrong leisure activities (such as fornication and debauchery) and be damned to the fires of Hell (Southern, 1990).

Scholasticism lost its grip to the new natural philosophy of the Scientific Revolution, though this process was slow, and the new philosophy owed much of its success to social and political factors, as well as epistemological reasons (Shapin, 1998). Two contested versions of the new natural philosophy emerged in the seventeenth century. In England, Bacon returned to a purer form of Aristotelianism, unencumbered by medieval theology. He argued for the primacy of observation and experience over tradition, and in his *The New Atlantis* of 1626 set out a utopian vision of the advancement of human well-being through knowledge (Bacon, 2008). In this utopia, men were free to gather the observations necessary for an inductive account of the world. From Bacon came the idea of the learned man at leisure, dabbling in scientific experiments and reporting finings to a democratic collective of philosophers, which was the inspiration for the Royal Society. This collective, of course, would then make judgements on the proper way to live, the right kinds of work for the masses, and the correct leisure forms that would encourage happiness and conformity. Against the bird-spotting instincts of Bacon was the rationalism of Descartes, which dominated Continental philosophy after a succession of best-selling publications by the Frenchman, including the *Meditations* of 1641 (Descartes, 2003). Descartes' lively style drew a picture of him dozing in taverns, contemplating meaning and realizing that the only certainty was his ability to think rationally. From this, Descartes deduced a mechanical world through which human minds – a combination of soul and body – moved. The power of the mind lessened the heresy of the mechanical world. In terms of leisure, Cartesianism promoted an instrumentalized version of the world, in which humankind could find satisfaction in hunting and baiting animals that were simply *automata*. The world was there to be exploited for our leisure, whether in the production of food and drink, or the creation of art (Bordo, 1987).

Descartes' meditations on the essence of humankind inspired others to find answers about the nature of humans and human society. Some, such as Hobbes in his 1651 *Leviathan*, were pessimistic about human nature (Hobbes, 2002). In *The Social Contract* of 1762, Rousseau argued that man (Sic.) in his primitive state is completely without morality – he does whatever he likes in his life, including his leisure (Rousseau, 2002). This savage state, says Rousseau, is not something that we should desire. We are more than our savage selves – we are children of reason. Morals are an expression of the Sovereign Will of the people in society. That will is part of the Social Contract we make when we make societies, the formal and tacit rules that govern every aspect of our behaviour from the power of the State to the choices we make in our leisure lives. Only direct democracy, says Rousseau – what Habermas (1989[1962]) identifies as communicative discourse in the public sphere – can protect that Sovereign Will from misuse.

Kant attempted to build a moral framework of everything, in a series of carefully argued essays and books. His contribution to ontology, epistemology and ethics was phenomenal for the time (Kant, 2007[1781]), and Kant continues to be one of the most important philosophers today. Kant was upset by the instrumental empiricism and scepticism of other Enlightenment philosopher such as Hume (Gardner, 1999). He wanted to demonstrate that the human mind did have some reasoning, critical faculty, which enabled right-thinking humans to find out the truth. For Kant, this critical faculty was responsible for deciding between truth and falsity, beauty and ugliness (Kant, 2007[1790]). It was, for Kant, the combination of our unique critical faculty and our bodily sensations that allowed us to be confident about our ontological, epistemological, ethical and aesthetical judgements (Guyer, 1992). These, of course, especially the latter two, influence our decisions about the meaning and purpose of leisure. In discussing free art, Kant says, 'we look on [it] as something that could only prove purposive (be a success) as play, that is an occupation which is agreeable on its own account' (Kant, 2007[1790], p. 133). Free art, of course, is a communicative leisure form. The value and nature of play are specified elsewhere, where Kant (Ibid., p. 159) writes:

> The changing play of sensations (which do not follow any preconceived plan) is always a source of gratification, because it promotes the feeling of health; and it is immaterial whether or not we experience delight in the object of this play or even the gratification itself when judged in the light of reason... We may divide this aforementioned play into games of chance, harmony and wit.

There are, then, for Kant, different types of playful leisure, but only those based on games of chance, harmony and wit give us gratification and, hence, moral and critical satisfaction.

In the nineteenth century, Kant's work inspired and incensed in equal measure. The idea of the numinous proved attractive to the philosophers and artists of the Romantic Movement (Berlin, 2000). Romanticism led to an engagement with nature and an embrace of the sublime in the wilderness (Stevens, 2004; Tang, 2008), a precursor of modern walking tourism. Romantic ideas of destiny and race combined in the philosophy of Hegel, who saw in History the inevitable progression from imperfect societies to the perfect folk of Prussia (Beiser, 2005). In his philosophical racism, the German race and its folk-culture were to be equally celebrated. This led to a turn to popular cultural and leisure forms in the new unified Germany, a construction of nation in the dance music and beer halls of Bavaria, and the physical education of the Prussian system. Other nations revived or created their own local leisure forms in the belief that History was their manifest destiny – from the martial arts of the East, to football and cricket in England. Philosophy gave nationalists an ethical argument for their causes, which created practical problems of establishing what was ethically correct in any given culture (Kymlicka, 2002).

Nietzsche was concerned with the problem of ethics, too. In a series of obscure and often contradictory essays, he tried to reconcile reason with nihilism (see Leiter, 2002). Many postmodernists (e.g., Rorty, 1989) argue that Nietzsche subscribed to a nihilist position: an extreme scepticism about everything, meaning that nothing can be known for certain, meaning that nothing can guide us in our choice of right and wrong. However, it is clear that Nietzsche did believe he had a set of guiding ethical principles about the good life – the natural, physiological supremacy of the strong; and the inequality of different types of humans – which led him to assume the moral correctness of hiking, exercise and reading (Hayman, 1980; Gibson, 1993).

In the twentieth century, philosophy was professionalized and institutionalized through the modern university system. Questions about the meaning and purpose of leisure, and the ethics of leisure, became relatively unimportant and ignored by a wave of analytical philosophers concerned with logic and semantics (Stroll, 2001). Philosophical debates about leisure would have to wait until the 1960s and the slow evolution of leisure studies and philosophy of sport (MacNamee, 2007; Rojek, 2010). One philosopher who did think about games and play was Wittgenstein. In his *Philosophical Investigations* (1968), Wittgenstein rejects his earlier formalism over the meaning of language. He argues that meaning is not something that can be learnt through pedagogical semantics (the point and label approach): instead, meaning is only ever learned through the use of words in language games. We learn the rules of each language game through playing those games, typified by this example (Ibid., §1):

> Now think of the following use of language: I send someone shopping. I give him a slip marked 'five red apples'. He takes the slip to the shopkeeper, who opens the drawer marked 'apples', then he looks up the word 'red' in a table and finds a colour sample opposite it; then he says the series of cardinal numbers – I assume that he knows them by heart – up to the word 'five' and for each number he takes an apple of the same colour as the sample out of the drawer. – It is in this and similar ways that one operates with words – 'But how does he know where and how he is to look up the word 'red' and what he is to do with the word 'five'?' – Well, I assume that he 'acts' as I have described. Explanations come to an end somewhere. – But what is the meaning of the word 'five'? – No such thing was in question here, only how the word 'five' is used.

In this thought experiment, Wittgenstein is demonstrating that in defining meaning and purpose one cannot find any essential truth, some meaning beyond the usage in any given language game. This has enormous implications for philosophy – and Wittgenstein's ideas have being attacked by many philosophers keen to keep hold of the semantic essentialism Wittgenstein rejects (Kuusela, 2008). However, Wittgenstein's work has proved immensely useful in helping sociologists and cultural theorists make arguments about the social construction of things previously perceived to be natural in some way (see, for example, Perloff, 1999). The concept of the language games in turn gives prominence to play, games and leisure in social discourse (Harris, 1990) – we can see the meaning and purpose of leisure in the social use of leisure. Wittgenstein's work is also a strong influence on philosophers of sport, play and games, as I will show next.

Contemporary debates in philosophy of leisure and sport

Contemporary debates in the philosophy of leisure have revolved around the differences and similarities between leisure as play, and more structured forms of leisure such as games and sports. In trying to expand on this initially problematic definition of leisure, Roberts and Parker developed an epistemological and ontological position on leisure that took it to be, in essence, something to do with free action, free will, free choice. This idea of leisure as freedom is one that a number of leisure studies researchers, not least Roberts himself (1999, 2000, 2004, 2011), have continued to defend. Ultimately, leisure and studies of it are to be predicated on a notion of a freedom to act sketched out by John Stuart Mill:

> The only part of the conduct of any one, for which he [sic] is amenable to society, is that which concerns others. In the part which merely concerns himself, his independence is, of right, absolute. Over himself, over his own body and mind, the individual is sovereign. (Mill, 1998[1859], p. 14)

On this understanding, leisure is part of a liberal capitalist industry that provides for our consumer needs (Bacon, 1997), and the task of researchers, as good empiricists, is simply one of following trends and explaining them. Choices are limited by circumstance and history, but there is a trend towards greater freedom of choice in leisure, especially evident in Western, neo-liberal economies (Roberts, 1999, 2004). Those choices are rational in both the classically economic and the Weberian sense. The discourse of leisure seen from this theoretical perspective is, historically, the discourse that most people in the past understood. By taking a liberal position on leisure, it becomes possible to understand the meaning of leisure in, for example, seventeenth-century Protestant fears of idleness (Weber, 2001[1930]).

Rawls (1971) inherits the traditions of Rousseau and liberals such as Mill, and attempts to establish a normative account of ethics and social justice. If an individual does not know how she will end up in her own conceived society, it is very probable that she is not going to privilege any one class of people, but rather develop a scheme of justice that treats all fairly. That system, says Rawls, would have two features. Firstly, each person would need to have an equal right to the most extensive scheme of equal basic liberties compatible with a similar scheme of liberties for others. Secondly, social and economic inequalities are to be arranged so that (Ibid., p. 303) (i) they are to be of the greatest benefit to the least-advantaged members of society (the difference principle); and (ii) offices and positions must be open to everyone under conditions of fair equality of opportunity. One can begin to see how such a normative philosophy of social justice – with its commitment to equality – might apply to thinking about freedom and ensuring that everybody has the time and resource to enjoy their leisure time without constraint.

Philosophers have also asked: what is a sport, and what is so unique about it? Sport is something to do with physical activity, which is more tightly defined than leisure, which can be physical or not (McFee, 2004). Sport is also defined as something to do with rules and rule-following: something to do with competition, strategy and tactics. In other words, sport is something to do with organization. As Fraleigh (1984, p. 41) puts it, a sport contest is:

> a voluntary agreed upon, human event, in which one or more human participants opposes at least one human other, to seek the mutual appraisal of relative abilities of all participants to move mass in space and time, by utilizing body moves, which exhibit motor skill, physiological and psychological endurance and socially approved tactics and strategies.

Sport is something that is associated with leisure, and play. But there is more to it. The role of the contest is the key to understanding its meaning. Fraleigh continues (1984, p. 41): 'The purpose of a sports contest is to provide equitable opportunity for the mutual contesting of the relative abilities of the participants ... within the confines prescribed by an agreed-upon set of rules'. Loy (1968) argues that a sport is just an institutionalized game. Suits (1988, p. 43) suggests, 'to play a game is to attempt a specific state of affairs, using only means permitted by the rules, where the rules prohibit use of more efficient means'. However, this definition is problematic. As McBride (1988) points out, although some games are sports (e.g., baseball), and some games are not sports (the card-game bridge), there is the complication of some sports not being games (such as fishing). Games could be conceptualized as activities that are bounded by rules understood by all participants. So, for example, a Knight in chess may be moved a certain way. If a participant moves the Knight a different way during a game, what happens then? The participant is reprimanded and obeys the rules. But another option is that the game continues with the agreement of the other participant: but the game then is no longer chess. Sports differ from games in having clearly defined rules and mechanisms for ensuring that rules are obeyed, and offenders are found out and punished. Games are morally more flexible, in one sense, because often there is no independent, objective court of appeal to refer to if participants feel 'cheated' by others. But, because of that, games often have a highly developed moral code (sometimes unwritten or tacit) that is policed by the participants themselves. So, for example, a game of hide-and-seek operates on an unspoken agreement that the person hiding does not catch a bus to the other side of town, but instead hides in an area familiar to the seekers. Similarly, seekers must give the hider a chance to hide, even if they do not always count to one hundred. Games are much closer to the playful form of leisure than more organized, more rule-bound sports.

What seems to be central to the argument of Morgan (2005) is that sport, by its ontological definition, is intrinsically something based on a fair contest: the concept of fair play is communicative in a Habermasian sense, that is, discussed and debated away from any conception of instrumental law, for example kicking the ball out of play when a player is injured. The problem with fair play is this: it is a normative ideal, but is it an actual intrinsic component of modern sport? Another approach might be to argue that sport's moral nature is associated with its rules. Rule-following in sports seems to be associated with the kind of social ethics Rousseau and Rawls talk about: the rules of sport are voluntarily agreed upon; there is no compulsion to play sport; but if you do play a particular sport, you abide by its rules (though you can take action to change those rules through membership of that sport's community). The problems with that, of course, are the unwritten or *tacit* rules that govern how the rulebooks are understood in various social and cultural practices. Some sociologists argue that philosophical definitions of fair play in sport are *idealistic*, and based on erroneous, tacit ideologies (such as breaking the speed limit). Pringle and Markula (2005) show how the sport of rugby allows permissible violence beyond the rules; there are tacit rules accepted by players, interpretations by referees, and the rules in rule book. This is to confuse meaning with purpose – the purpose of the ontology of rule-following may be to normalize ideologies; but the meaning of the ontology of rule-following is still clear; we can be normative about rule-following, even if we are aware of the way the normative *ceteris paribus* conditions work badly in practice.

Classic functionalism argues that sport is an instrument in the rationalization of society (Weber, 1992[1922]). Figurationalism (Elias and Dunning, 1986) argues that modern sport represents a controlled place where emotions can be exercised in a civilized world, and hence sport is part of a wider civilizing process. Social philosophers such as Morgan (2005) argue that sport can be a societal good, as long as it is allowed to be true to its ontological self – when it is not, it can no longer be good for society. But what is its true ontological self? Can sport be said to have one, essential nature? Foucault (1970) argued against making any such assumptions about any social or cultural activity (which sport clearly is); all these things are contested, and their meaning and purpose subject to struggles over power and control. Gibson (1993) used the work of Nietzsche to explore the purpose of sport. Nietzsche (reference) claimed the Enlightenment was over, and rationality was to be replaced by emotion and struggle. In the new world, free will would be replaced by the will to power – a Darwinist struggle of might that would see the emergence of a new race of *ubermensch*. In modern, professionalized sport, argues Gibson, the purpose of sport has become to test and support struggles between such 'supermen' – the Olympics then becomes a realization of Nietzsche's nightmares of the modern.

Huizinga (2003[1944]) distinguishes play by its 'non-serious' nature. Play is seemingly frivolous activity that is not, ostensibly, about any of Maslow's (1998) higher needs. Sport is codified play, active leisure. The motivation to participate in leisure can be viewed as a way of meeting psychological, social and cultural needs. This leads to other qualities Huizinga identifies: play is free; is self-contained; is regulated or rule-governed; is limited in space and time; is 'make-believe'; and is tension. Play can be understood, following Huizinga, as a quality (that is, intrinsically motivated, for fun). Play, and games, and sports, become important for the value they have in being free, non-work: the freedom to game-play, which Suits (2005) sees as the essential quality of his famous version of Utopia. Play can also be understood by exploring its outcomes – that is, exploring how it is related to extrinsic motivation and function. It can be understood as developmental and educational (Cohen, 2006). Play can be described as stimulus-seeking behaviour (Hughes, 2001). Mead (2001) suggests play has a role in socialization: developing a sense of self and Other defusing conflict and enculturation. Play can also be seen as compensation and sublimation (Singh, 2001). The greater the freedom allowed to players to choose an activity, then, the greater that activity's potential as leisure. The more extrinsically orientated a leisure activity becomes, then the lower its leisure potential (Hughes, 2001; Cohen, 2006). The better any activity fits with qualities of play, the greater its functional significance. The greater the emphasis on play as a vehicle for intentional learning, the lower its leisure potential becomes. The more structured a play form becomes, the less its leisure potential – and the greater the degree of organization in a game, the lower the degree of freedom and choice available to players. For example, you cannot pick up a football and run with it, despite the myth of William Webb Ellis inventing rugby in 1823 (Collins, 1999).

Holowchak (2007) examines the main philosophical objections to Suits' (2005) vision of a Utopia defined by game playing. He acknowledges that Suits' argument provides an incoherent account and insufficient confidence in the assumption that the relationship between Utopia and game playing holds beyond some wishful thinking on Suit's part. In recognizing this, he shows that Suits' utopian leisure is based on an over-confident acceptance by Suits of a Platonic turn. However, Holowchak (2007, p. 95) then claims that there is evidence that Suit's Utopia may be closer than we think, realized by advances in technology and societal change:

> Utopias of a deeply fulfilling sort may be much easier to attain than the sort Suits envisages. Perhaps, as Plato reckons in *The Republic*, the things we really need are few and relatively easy to acquire. Only when our wants outstrip our needs do we seek out largely 'inefficient' means to satisfy ourselves. We build luxurious domiciles, trade for perfumes and fancy linens, cavort with paid escorts, and even begin to war with other peoples, just so we get the things we want but do not need. Only when we abandon Plato's vision for a simple lifestyle, embodied by his *kallipolis*, do we seek out the sort of prodigal, high-tech society that Suits envisages in his Utopia.

Morgan (2008), also commenting on Suits, is, however, much more sympathetic with the matter-of-factness of Suits' use of concepts like games and play. In his defence of Suits, he draws attention to the work of Wittgenstein (1968) and writes:

> We get ourselves into such absurd definitional binds, to reiterate Suit's caveat, when we accept what people call play (any autotelic activity whatever), as the gospel truth, as an adequate factual description of play. To avoid designating a cat chasing its tail a religious experience and Aristotle contemplating God a playful one, we must, he insists, first establish whether the facts warrant such a classification. And to do that, Suits continues, we must steadfastly reject the advice Wittgenstein's offered in a famous passage in *Philosophical Investigations*, in which he tells us, focusing ironically enough on games, 'Don't say 'there must be something common or they would not be called games,' but look and see whether there is anything common to all' (quoted in Suits, 1977, p. 163). We should shun Wittgenstein's advice, according to Suits, because the question we need to ask here is not whether all things *called* games share something in common, but rather whether all things that *are* games share something in common (Suits, 2005).

Afterword

Aristotle provided a problem that has constantly vexed philosophers up to the present time, so that postmodernists are rejecting the Aristotelian Truth, by drawing on inductive logic and comparing it to deductive logic, and coming up with nothing, whilst seekers for the Truth have struggled to reconcile beliefs, ideas and experience with the neat solutions provided by syllogisms and their internal truth value. Someone once said everyone in the world of philosophy is either a Platonist or an Aristotelian. I suggest that we should all learn how to be Platonists. Plato described truth, and reality, in a different way to Aristotle, and it is this break between master and rebellious pupil that laid the ground for the history of ideas. For while Aristotle came to be so influential in providing us with our true Truth, our deductive logic and our epistemological problems, Plato was less influential on modern philosophy. At periods through history, his work has influenced philosophy, mainly through scepticism, mystic hermeticism and the dividing of reality into the Forms – the really real things of which the divine and the soul were a part – and the poor, corrupt imitations of the Forms that were physically here, for example, the world out there, and you, the reader, holding this book. It is this final idea (the idea of Ideas, as the Forms are sometimes known) that relates to my argument about leisure.

Essentially, Plato said that there were two types of truth: *episteme*, which related to the Forms, and hence was really true and could be proven; and *doxa*, which was the kind of truth that could be related to the changeable, corrupt world, that gave rise to opinions, and which was not really True, in a sense of the only Truth being part of the Forms, to which *episteme* was directed. There was a truth hierarchy. At the top was a meta-truth, that of the Forms, and below that was the *episteme* status of truths that related to that meta-truth. Finally, there was *doxa* truth, which related to things like experience, discourse and what we would like to call science. So, knowledge could be had either in an epistemic or doxic sense. We can begin to understand the search for the meaning and purpose of leisure as a search for doxic and epistemic explanations of leisure, and doxic and epistemic purposes of leisure. So, accounts of leisure through history turn from the everyday to the metaphysical, the communicative to the hidden power of instrumentality. I will return to these ideas in the penultimate chapter of this book, and show how they might help us understand how we can understand the role of leisure.

3
Leisure and Human Nature

My historiography begins in this chapter with an exploration of leisure in pre-historical times and pre-literate – or essentially oral – cultures. This will explore archaeological evidence from a range of international settings. The first section of the chapter will concentrate on the pre-history of humanity in general, with an account of the deep history of leisure in the cultural lives of *Homo sapiens* and our earlier hominid ancestors. I will argue that archaeological, physiological and psychological evidence, along with comparative work with modern-day primates, suggests that leisure has always had a pivotal place in the lives of humans. In the second section of this chapter, I will concentrate on the material evidence for leisure in the lives of humans in the Neolithic and Bronze Ages, using finds ranging from combs to cave-paintings to identify increasingly complex leisure-choice rationality. The third section of this chapter will examine mainly oral cultures such as the Incas (see, for example, Smith and Schreiber, 2006) and other cultures of the Americas. I will argue that the political and social complexity of those Native American and Amerindian cultures has been probably distorted – partly by Western historiography and exchange in the historical period, but also partly by the elites of those cultures who transmitted hegemonic ideas of their status to Westerners during that period of exchange. The fourth section of this chapter will continue this theme through discussing the leisure lives of 'primitive' cultures observed by early ethnographic anthropologists of the nineteenth and twentieth centuries. I will argue that such ethnographies were laden with assumptions about the irrational nature of their subjects, and failed to provide a true account of agency, change and debate in those cultures. The chapter will then end with a critical discussion about the gap between material remains and the meanings ascribed to them by archaeologists (Courbin, 1988).

Prehistory and the development of humanity

The 17,000-year-old Palaeolithic artwork on the caves of Lascaux, France, was famously discovered in 1940 by a gang of local teenagers. Leroi-Gourhan (1968) introduced the work to a wider audience through the publication in English of *The Art of Prehistoric Man in Ancient Europe*. This book was published as a modern art book, with glossy photographs of the images alongside Leroi-Gourhan's interpretive and critical text. The artwork itself, created using mineral pigments and some carving of rock surfaces, consists of approximately 2000 images: abstract patterns, clearly defined animals, and representations of humans. Leroi-Gourhan argues that the artwork demonstrates a high degree of intellectual and psychological capacity among its maker(s), in other words, the cave paintings should be considered as works of art, of human genius.

These cave paintings and others in France and Spain have led some archaeologists and anthropologists to claim that they marked the arrival of modern humans into Europe (Bender, 1978). The paintings represent the evolution of the human mind and the success of modern humans (Humphrey, 1998). For Conkey (1989), cave paintings of the kind at Lascaux are evidence of a cultural revolution: evidence in the mythological symbolism of the paintings (Leroi-Gourhan, 1968; Rodrigue, 1992), or the complex social systems that must have supported the makers of the paintings (Bender, 1978). Lewis-Williams and Dowson (1988) even suggest that the stylistic animals and repeating abstract forms are evidence of the use of psychoactive substances, and hypothesize controlled rituals where the artwork was consumed alongside such food and drink. What this debate about function and purpose reveals for historians and philosophers of leisure is that here, in Lascaux, 17,000 years ago, humans invested time and energy into creating something that came out of their imagination, something that they conjured up out of time spent thinking of something other than preparation of food and shelter. And in making these pieces of art, the makers allowed others to spend some time and energy, as the archaeological remains suggest (Stevens, 1975), visiting the caves and looking at the paintings. Of course, this may have been driven by some religious imperative – but as Rodrigue (1992) shows, spirituality seems to arrive in pre-historical human societies after the transformation to a life of sedentism or local transhumanism. What this shows, however, is that the humans who lived in and around Lascaux had leisure time and used it to create something which they valued, something which others in the same place appreciated.

The scientific examination of the evolution of humanity suggests that the last 6000 years have been a mere moment in the long and deep history of our *Homo* genus and *Homo sapiens sapiens* species. The Biblical narrative of the creation of the world in six days, Adam and Eve in the Garden of Eden, and the monotonous iteration of son after son after son in the Old Testament, led to Christian theologians throughout the history of the Church (from Eusebius to modern-day Creationists) arguing that the world – and all species, including human beings – was only six thousand years old. Advances in scientific knowledge in geology, physics, archaeology and biology in the nineteenth century demonstrated that the world was far older than the Christian mythology suggested – in 1859's *On the Origin of Species*, Darwin (2009) suggested that all species, including humans, had evolved from earlier species through a process of natural selection. There is not room in this book to discuss the social factors behind the acceptance of Darwin's theory of natural selection, but there are some notable points to make clear. Debates about the purpose of science in society (Desmond and Moore, 1991) and the development of science as a profession (Turner, 1993) were important to the acceptance of Darwin's long argument on natural selection. However, Darwinism owed its success as a theory to two things – the construction of an imagined community (Anderson, 1983) of evolutionary scientists (Desmond, 1989) to make the case for the theory, and the theory's appeal to the rules of scientific method laid down by Darwin's predecessors, in particular, Whewell (Depew and Weber, 1995). Darwin aimed to seek a consilience of inductions (Ruse, 1979), that is, a theory that provided a true, realist explanation of a number of seemingly unrelated facts, such as the appearance of geological strata, the variation of animals, the existence of fossil remains and other examples. As Thompson argues, Darwin's theory of evolution can be best described as a hierarchy of models in the sense used by proponents of the semantic view (Thompson, 1983).

Whether Darwin believed in a God or not (Anglican or otherwise) is a contested fact (see, for example, Brown, 1985) but not relevant to this discussion. What is important is that defenders and opponents of the theory of natural selection split across the camps of believers and liberal agnostics alike (Desmond and Moore, 1991). Although the fiercest cheerleader for Darwin, Tom Huxley, used the theory of natural selection to browbeat Anglicans, the relationship between religion and evolution was more complex than the caricature promoted by popular histories such as Hazelwood (2001). Huxley's contribution to the emerging science of man, 1863's *Man's Place in Nature*, was the first systematic attempt to use Darwin's theory of natural selection to account for the evolution and development of man. Huxley's application of the theory of natural selection to man and his place in nature attempted to show that there was a continuity and similarity between man and apes, and that this continuity was due to a physical process of causation: Darwin's theory of evolution, which would explain the process either as modification from a man-like ape or evolution from a common ancestor to both man and ape (Huxley, 2001, p. 108).

With this wider understanding of the biological mechanisms and processes that led to the evolution of modern humanity came evidence from archaeology of our earliest ancestors. Throughout the twentieth century, pre-history was carefully examined through the archaeological record, and various scientific tests have established a rough chronology of our evolution (Renfrew, 2008). Put simply, the earliest identifiable *Homo* species, *Homo habilis*, existed in Africa around two million years ago. From this species evolved both *Homo erectus*, a species that colonized other parts of the globe in the period from two million years ago to half a million years ago, and *Homo ergaster*, which appeared at the same time and from which *Homo sapiens* and *Homo neanderthalensis* derived no earlier than (but possibly much later) half a million years ago (Canfield, 2007).

One of the key debates in the evolution of humans is the influence of material and cultural practices on the development of the mind (Renfrew, 2008). Something happened in the evolution of us to make us, modern-day humans, what we are today – intellectually, physiologically and psychologically. In the deep history of human evolution, there are moments of gradual change, such as the emergence of *Homo sapiens* and the decline of the competitor species *Homo neanderthalensis* (Canfield, 2007), alongside more radical changes such as the switch from a raw diet to a cooked diet that evinces the discovery and use of fire (Gosden and Hather, 1999). Through time, the use of tools becomes increasingly complex, indicating more sophisticated ways of food preparation, the use of clothing and other material goods lost to the archaeological records, and the importance of grooming (Gamble, 2007).

Homo erectus is associated with highly adapted tool-making finds (Renfrew, 2008), but it is only in the last 50,000 years that such tools become associated with burials, as well as beads and other material adornments, that suggest an awareness of cultural signifiers and taste (Lewis-Williams, 2004). In the evolution of modern humans, then, there is a turn to culture that comes while the material finds still indicate transhumanism and hunter-gathering (Gamble, 2007). The physical act of hunting and gathering to survive is one that *Homo sapiens* inherits from *Homo erectus*, yet the meaning of such work is changed by the cultural significances shaped by our earliest ancestors (and their competitors the Neanderthals). At some point, humans invested a significant amount of time in dressing the dead in finery, and preparing their graves with goods for eating, playing and hunting (Canfield, 2007). This tells us a number of things. Firstly, humans felt some psychological connection with the dead. Secondly, they believed that the goods had to be buried with the dead for some reason – whether to symbolically 'close' the relationship between the living and the dead, or to give the dead tools to navigate their way through some mind of afterlife. Thirdly, the grave goods were not all associated with work: they show an awareness of cultural life and leisure activity.

Between the work of life, then, *Homo sapiens* and *Homo neanderthalenis* were beginning to display concerns with each other that went beyond the physical bonding of family or kin. They were learning abstract, cultural concepts such as status, desire, spirituality and power, as well as the importance of leisure. This cultural turn may have something to do with the development of language (Canfield, 2007), which may have some bearing on the development of modern humans and their large brains, although it has been argued that the cultural turn appeared thousands of years before the physiological adaption of vocal chords necessary for speech (MacNeilage, 2010). We can see that modern-day primates, without language, still find place for leisure and culture in their daily lives (McGrew, 2004): they idle around, indulge in play-fights, use tools, groom each other, and they are keenly aware of status. Perhaps, then, an increasingly effective use of fire, along with environmental factors that led to a greater abundance of food, amplified those primate-like leisure and cultural activities in our ancestors, which led in turn to the evolution of speech and other physiological adaptations that gave us an evolutionary advantage over our environment (Renfrew, 2008). This would account for the huge cultural success of the humans at Lascaux, and the impact of leisure on the lives of those who created the art, and those who supported its creation. A psychological awareness of leisure, then, and its benefits in terms of intellectual and cultural development, probably marked out *Homo sapiens* and *Homo neanderthalensis* from other hominids and primates. The old archaeological stereotype of 'man the hunter' is replaced by the human being enjoying its leisure time, alone or in the company of others. The archaeological, physiological and psychological evidence, then, along with comparative work with modern-day primates, suggests that leisure has always had a pivotal place in the lives of humans.

Oral cultures in prehistory – Europe and the Near East

In 1989, archaeologists excavating a Neolithic settlement in the modern country of Jordan found a rectangular stone modified by human intervention (Rollefson, 1992). Two regular rows of depressions worked out of the stone's surface gave clear indication of significance to the stone. The stone was dated through its location in the settlement sections to approximately 6000 BCE. In comparing the find to identifiable game boards for the traditional Arab games *Mancala* and *Wari*, Rollefson (1992) has claimed the find is a rare example of a find that demonstrates the social complexity of Neolithic prehistory. He argues in his findings in the *Bulletin of the American Schools of Oriental Research* (Rollefson, 1992, p. 1):

> Competition for the sake of entertainment pervades the fabric of all known societies and is documented virtually throughout the entire historical record. It is probable that this typically modern human trait has characterized social intercourse at least throughout the evolution of *Homo sapiens sapiens*... [it is rare to find] extraordinary (i.e., non-utilitarian) aspects of life in the prehistorical period... It [the board] verifies that Neolithic people had leisure time to win or lose at games of chance or skill during a 'revolutionary' time of human development.

There is, in this discovery, and others like it (see Whittle, 2003), increasing evidence in the material record of Neolithic engagement with leisure activities and pastimes. The Neolithic revolution in culture – at least in the Near East and later Europe – saw the first use of agriculture, and the establishment of settlements that became towns (Gimbutas, 1963). There is a debate among pre-historians about the causal relationship between settlement and agriculture (Whittle, 2003; Renfrew, 2008), but whichever came first, they are both evidence that humans in the Neolithic were becoming more efficient in the use of time and skills. Specialization in work and production was beginning to emerge with the settlements: potteries, for example, and areas of flint-axe creation (Whittle, 2003). In the Near East, in what is now Turkey and Iraq, complex Neolithic town-based cultures appear in the material record. At Catalhoyuk in Turkey (Balter, 2005), there is evidence in the remains of the 9500-year-old town of a society concerned with the creation of private, domestic spaces (with evidence of hearths alongside material of a more uncertain nature – Sagona and Zimansky, 2009) alongside more spacious buildings that were undoubtedly for public use. On the northern edge of Europe, at Scara Brae in the Orkneys, a small village built of stone was nearly all private in its use of interconnected small buildings. Here, too, there is evidence of domesticity, but also perhaps of leisure in the finds of shelving spaces (Cowan, 1989).

As the Neolithic becomes more sedentary, there is an abundance of material evidence to suggest that the farmers and specialized workers in the small towns of the Near East were living relatively contented leisure lives. Surplus food from harvests was traded for specialities from other sites, often hundreds of miles away. People dined on exotic foodstuffs (Gosden and Hather, 1999), drank wine and beer (McGovern, 2011), used make-up and combed their hair (Whittle, 2003) and played musical instruments (Sachs, 1938). Settlement supported intellectual evolution and the creation of written languages (Whittle, 2003). These small towns were the forerunners of the empires we know from the inscriptions and fragments of writing left to us from the first literate peoples (Robinson, 2007), and the references to these people in the work of Greeks such as Herodotus and Manetho: the Assyrians, Babylonians, Hittites and Egyptians. Elsewhere, the Neolithic cultural revolution was slow to change the habits of humans. In areas of poor soil and climate, hunter-gathering remained the norm (Renfrew, 2008). Elsewhere, such as northern Europe, transhumanism and pastoralism emerged as compromises towards a fully agrarian economy. Both of these ways of food production, however, allowed small settlements to appear, with some permanency of inhabitation, and some free time for humans to think about filling that time with diverting or morally correct activities. In North Africa, on the fringes of the Near East hotbed of the Neolithic, the cave-paintings of the mountains of the Sahara desert indicate a transhumanist culture rich in abstract thought (Shaw, 1936).

By the Bronze Age, agricultural economies were establishing large empires, inequalities of wealth, and cultural and linguistic hegemonies (Kohl, 2009). Egypt is the most famous example in the Near East, with a society that allowed the rich to give their children gold toys to play with, while the lower orders were routinely denied their leisure time as their own to be enrolled in wars, the construction of civic works or the compulsory attendance at religious rites (Bard, 2007). Such hegemonic instrumentality of Bronze Age societies must have allowed the construction of sites in Europe such as Stonehenge (Bradley, 1998): leaders rich on surplus wealth expecting tributes of tithe and time from their subjects. However, the material wealth of the Bronze Age and the later Iron Age, and the imperishability of the metals that mark these pre-historical periods in Europe and the Near East, have led to an abundance of material evidence of leisure. The rich were surrounded by opulence – slaves, drink, fine food (Bard, 2007) – but the common humanity of these times is found in the evidence of beer-drinking (McGovern, 2011), gaming counters (Bard, 2007), cauldrons (Gosden and Hather, 1999) and bodily adornments (Kohl, 2009).

This material evidence for leisure in the lives of humans in the Neolithic and Bronze Ages, using finds ranging from combs to cave-paintings, allows us to identify increasingly complex leisure-choice rationality among people who lived in these times. They were, where possible, expressing their lives in a communicative use of their leisure. But all too often, and increasingly in the Bronze and Iron Ages, they were subject to the instrumental desires of despots.

Oral cultures in prehistory – the Americas

Archaeologists and historians have identified a series of long-lived cultures in the Americas, which flourished before the arrival of European colonizers. The most famous of these cultures were civilizations based on imperial hegemonies and cities: the Incas in Peru (Rostworowski, 1998), and the Mayans and Aztecs in Mexico (Smith, 2002; Hendon and Joyce, 2003). The Incas and the Aztecs are famous in Western European histories because they were empires that met the full force of Spanish colonization, defeated then destroyed by deceit, disease and superior European weaponry (Hart, 2008). Others left fewer stone-built material remains and remained distant from the first battle-grounds of the *conquistadores*: the plains-dwelling Native Americans (Scheiber and Mitchell, 2010), or the Inuit of the Canadian Arctic (Stern and Stevenson, 2006). These cultures did not necessarily flourish at the same time: different cultures and civilizations were succeeded by other cultures (through force or economic transfer), and some sank into obscurity to be revived later under different names.

Aztec was a later name for the indigenous, Nahuatl-speaking population of part of what is now called Mexico (Smith, 2002). In the thirteenth century, the Aztecs were concentrated in the valley of Mexico: the people of the Mexica tribe. The Aztec Empire was an alliance of three city-states in the fifteenth century CE, centred on the city of Tenochtitlan built on Lake Texcoco. By 1521, the empire stretched from coast to coast of modern-day Mexico. The Aztecs, like their Mayan predecessors, are known for their ball games, which were first recorded by Spanish colonizers, and which are evinced by the proliferation of game sites in Central America and a number of inscriptions. Cohodas (1975), drawing on the archaeological work of Brom (1932), describes the ball-game of the Aztecs in a fulsome, journalistic manner. He writes (Cohodas, 1975, p. 99):

> Of all the ball games that were being played ... at the time of European contact, the Mesoamerican game, which used a rubber ball and was played in masonry ball courts, was the most elaborate ... The ball game was clearly an exciting spectator sport, emphasizing both stamina and skill.

The ball game was a team sport and was popular in terms of participation and spectating. But it was not just a spectator sport where athletes were cheered on. The game had a meaning and use among Mayans and Aztecs far removed from the modern professional team sports such as football. The ball games, like all aspects of Aztec everyday life, were highly instrumentalized in a culture where politics and religion were intimately connected. The popular belief in archaeology and anthropology that these games were simple equivalents of professional sports of the twentieth century CE is based on a simplistic reading of the evidence. As Fox (1995, p. 485) describes:

> The assigning of ballgames to archaeological ballcourts is based primarily on ethnohistoric and iconographic studies. Spanish descriptions of Aztec ballgames taking place in masonry ballcourts, along with Aztec and Classic Maya imagery showing ball playing in progress, were early linked to the archaeological remains of ballcourts being encountered in the Maya Lowlands (Blom 1932). From that point on, the architectural morphology of ballcourts became the focus of archaeological study as the one material link to the ancient activity of ball playing.

The Aztecs were deeply religious, believing in a number of gods who played active roles in Aztec life (Smith, 2002). Aztec religion and mythology inspired the building of massive stone temples and sacred sites, normalized human sacrifice and supported the aristocratic warrior class (Hendon and Joyce, 2003). Every 52 years, Quetzalcoatl, god of lightning, was said to take some sort of active role in Aztec affairs (Smith, 2002). The leader of the Spanish conquest of the Aztecs, Cortes, left Cuba in February 1519, landed at Cozumel, then sailed around the Yucatan to Potonchan. He founded the colony of La Villa Rica de la Vera

Cruz. He seemingly encouraged the Aztecs to believe he was Quetzacoatl, though the impact of this (and whether the Aztecs believed it) is much disputed (Restall, 2004). He made an alliance with the Tlazcalteca, enemies of the Aztecs. In November 1519, Cortes and his men reached Tenochtitlan and were greeted as guests by Moctezuma, who was then imprisoned by Cortes and killed (Hart, 2008). The Aztecs rose up against the Spanish, and Cortes fled with his men and his allies, but he soon returned: in 1521, he besieged Tenochtitlan, and in August of that year the city surrendered along with the last Aztec emperor (Smith, 2002). This concern with ritual and power is, according to Fox (1995), evidenced in the multiple meanings and uses of ball courts in Mayan and Aztec sites. The spaces of the ball courts were not merely places of recreation: they were places where religious rites, sacrifices and offerings to gods and emperors were made.

In South America, the Incas also developed a highly centralized empire, which, in turn, was invaded and defeated by the Spanish soon after the fall of the Aztecs (Rostworowski, 1998; Hart, 2008). Some of the early Spanish sources on the Incas mention the use of coca, a plant-leaf with narcotic properties, which they argue was reserved for the elite classes for ritualistic purposes (Bolton, 1976; Rostworowski, 1998). This apparent edict – the Incan limitation on the use of coca – has become a story that appears as fact in a number of academic sources on the ethics of drug use (see Gootenberg, 1999). However, there is strong archaeological evidence to suggest that the use of coca was widespread across the Incan territories and across all Incan classes (Bray and Dollery, 1983). This is seen in the survival of coca in archaeological sites, the representation of coca leaves in decorative artwork, the survival of coca pouches, and indirectly in the widespread use of coca in the Andean region today (Bolton, 1976). Put simply, coca was the drug of choice, the leisure stimulant of choice, for the everyday Incan in the street. There is evidence that the Incan state controlled the production of coca, and coca leaves were used as a quasi-currency in Incan times, but coca was freely available and used by all free Incan men and women (Gootenberg, 1999). Coca was not simply a narcotic – it is wrong to see it as equivalent to opium in nineteenth-century CE China, or crack in twenty-first century CE America. Coca as a stimulant in the Incan culture was probably closer to coffee in the Ottoman Empire, or tea in twentieth-century CE Great Britain: it was an uncontroversial accompaniment to social exchange and something deemed good for helping one get through the working day.

Why the Spanish recorded the story of aristocratic prohibition is unclear. It may have been that the rulers of the day had turned against coca use among the masses, for some reason associated with the pressure of the Spanish arrival and the attempted takeover of Incan life. Like the ball games witnessed by the Spanish when they conquered the Aztecs, the complex leisure life of the Incas has been frozen in the writing of European colonizers. This colonialist lens is apparent in the way in which the Native Americans of North America have had their leisure activities portrayed. Much of what people think about the Native Americans is refracted through that lens in a way that either: idealizes such leisure and culture as being a pure, unsullied ecological and spiritual one; or demonizes the native Americans as savage primitives, the Red Indians of a thousand comic strips. The cultural diversity and complexity of actual lived leisure lives of Native Americans are hardly ever mentioned in Western films and comics: the importance of smoking in male-bonding rituals (Winter, 2000); craftwork as resistance (Maruyama et al., 2008); and the introduction of alcohol along the route of the railroads, and its deleterious impact on Native American society. Frank et al. (2000, p. 348), in a study exploring at the history of alcohol abuse among Native Americans from pre-Columbian times to the present, recognize the way in which Native Americans learned to drink from the example of the white Europeans they first met:

> Drinking was pervasive among the early European colonists, and alcohol served practical purposes in their everyday lives. Alcohol was used as a substitute for drinking water (which was usually contaminated) and for medical purposes – to fight fatigue, soothe indigestion, ward off fever, and relieve aches and pains.

There was some complicity in the use and dissemination of alcohol – individual Native Americans liked the taste, or the intoxication, and many chiefs saw the drink as a way of reaffirming their own power through control of its use and distribution. People were capable of using their own agency to make Habermasian, communicative decisions to drink or not to drink. Chiefs saw Habermasian instrumentality in their manipulation of alcohol supply and their manipulation of the white men who sold it. However, Frank et al. (2000, p. 349) also note that the introduction of alcohol into Native American culture was often a deliberate policy of disruption and disorientation by the Western powers, aside from an act of capitalist greed:

> Few commentators have compared the Opium Wars with the colonial powers' use of alcohol in their dealings with Native North Americans.

> In both instances, more technologically advanced peoples deliberately planned and executed the habituation of traditional peoples to a damaging psychotropic substance for the purposes of economic benefit and territorial expansion.

The political and social complexity of those Native American and Amerindian cultures has been probably distorted – partly by Western historiography and exchange in the historical period, but also partly by the elites of those cultures who transmitted hegemonic ideas of their status to Westerners during that period of exchange. But the fact remains that a consequence of the history of European colonization in the Americas was the destruction of native cultures and disappearance of many traditions, myths and leisure activities – and the loss of countless lives.

The lens of imperialist anthropologies

Anthropology as an academic discipline has a troubled relationship with Western imperialism and Western hegemony (Tilley and Gordon, 2007). The first anthropologists of the nineteenth century CE were trained in Western universities, or were Westerners dabbling in anthropology as rich gentlemen-amateurs or interlopers (merchants, engineers, sociologists, geographers, soldiers, officials) in the non-Western cultures they described (Barnard, 2000). They were interested in exploring and describing cultures that lived beyond the boundaries of Western civilization, or in its darkest depths. These cultures were supposed to represent some kind of pristine, antediluvian state of being, the primitive savage of nature idealized in the moral philosophy of Rousseau. Anthropology set out to try to map and explain the nature of 'man', from a starting point that assumed, in a naïve way, that industrialized nation-states had fallen from some sort of grace and had become less authentic as examples of human culture. Sociology, said the anthropologists, was the science of modern society – but to understand the nature of human culture, it was necessary to find some distant place where Western civilization (sports, trains, telegraph poles, trading stations) had not reached (Brownell, 2009).

Anthropologists established the methodological practice and rule of ethnography (Marcus and Fischer, 1986). It was good anthropological science to examine the native cultures in these exotic places from as many angles as possible, but it was best practice to live with among the peoples being investigated, to get to know their ways, their customs, their norms and values, to get to the heart of their realities, their lives (Barnard, 2000). Spending time, digging deep, this was the tool of ethnography, which allowed anthropologists to return to the West to write up their research. This led to a string of important anthropological studies, published by Boas (1988[1928]) and others (such as Durkheim and Mauss, 1969; Mauss, 1990), which demonstrated, very swiftly, the strange diversity of human experiences. Rather than finding identical hunter-gathering communities, all with primitive belief-systems and tribal structures, anthropologists found a rich set of data that seemed to suggest that humans were infinitely adaptable and infinitively creative. Rather than a consensus of meaning, anthropologists found cultural relativism, where people believed in an enormous range of gods, spirits, worlds and systems (Turner, 1982). If one tribe believed that the dead should be cremated to appease a wrathful fire-god, and another believed that cremation was the worst sort of sin with fatal ethical consequences, how could anthropologists develop a science of human culture? That was one problem to be solved; another was the observer-bias that anthropologists imposed, and the hegemonic nature of their own, locally created Westernness in its interaction with subaltern cultures. Such ethnographies were laden with assumptions about the irrational nature of their subjects and failed to provide a true account of agency, change and debate in those cultures.

The answer was to develop systems, structures or theoretical frameworks that would allow anthropologists to compare different sets of data to try to find some meta-language or meta-narrative that operated at a higher ontological level that the culturally relative facts gathered in the field. This would also help remove bias and subjectivity from the discipline. Twentieth-century anthropologists tried to do this with varying degrees of success. Levi-Strauss (1963) offered a structural synthesis approach, which found common structures between different cultures. Mead (2001) reified cultural relativism itself, allowing a plurality of voices and a more equal engagement between her subjects and herself as author-Westerner. Geertz (1973) took a symbolic line, arguing that human cultures were spun out of the webs of significance – the symbols, the meanings – each one of us makes when we interact with others. More recent anthropological theory has drawn on postmodernism to argue there are no truths to be found, no structures to help make meaning (Barnard, 2000), or to argue that the science of anthropology is a handmaiden to racism and Western control of the rest of the world (Brownell, 2009). However, such epistemological relativism, while seemingly reflecting the fact of cultural relativism inherent in the plurality of data, is self-contradictory as well as self-denying. Anthropology has served to foster racism and Western prejudices about the reified Other, but it has also shown that only a universal system of logic can recognize this and the sameness of Western culture compared with every other culture in the world.

What this debate leaves this study is the confidence to explore and understand the human condition in all its local particularity. There are very few anthropological universals. But one thing is clear from any attempt to synthesize an account of what it means to be human. In all cultures studied, the anthropological evidence indicates that people value activities that are not simply oriented towards the fulfilment of basic needs such as food and shelter (Barnard, 2000). People from all cultures have a desire to enjoy the company of other people, to eat in social gatherings (Mead, 2001), to listen to stories, poetry or music (Geertz, 1973). Children everywhere play games, shaped from their specific cultural backdrops and material and environmental circumstances (Monaghan and Just, 2001). People value the time spent engaging in these activities, and the time spent hanging around doing nothing much at all of seeming value (Geertz, 1973). There are exceptions, of course, to all this, in the differential freedoms and practices associated with different castes, groups, classes, ages, sexualities and genders. What this differential demonstrates is the way in which cultural norms are constructed and used within those cultures by those who have access to the instrumental rationality associated with power and control. So, it is in the interests of men to limit the leisure activities of women to domestic spaces and say women are unclean if they join men in a public leisure space (Levi-Strauss, 1963). But there is an essential truth to the human condition, as described by anthropologists such as Geertz (1973) and Mead (2001): it is human to value one's leisure time, and human to try to control the leisure of others. This may have come from a discipline associated with the lens of imperial hegemony, and it may be a product of liberal beliefs about leisure imposed on the data by biased anthropologists, but it is still undoubtedly truth of sorts.

Material objects and meaning

Throughout this chapter, I have referred to evidence from the archaeological record: essentially, material remains and objects found *in situ* by archaeologists, or artefacts stored in museums that may not have a known site of origin. None of these objects comes with a label attached to them that says, 'a gaming counter used by Babylonians', or 'Aztec ball court'. Archaeologists have to rely on prior knowledge (e.g., the pottery sequences that help establish relative dating) and relationships to interpret the meaning of these objects. Sometimes, this may be a straightforward problem to solve: a Viking antler-bone 'comb' may be proven to be a comb by finding in its teeth the remains of head-lice; it may be proven to be Viking by finding it in the remains of a site in Sweden proven to be of Viking-era date through carbon-14 dating or coin finds. However, in this discussion about the leisure lives of people in pre-history, or people in oral cultures, we are faced with a more serious problem of moving from the material facts of the data to the meaning and purpose we interpret.

Some theorists have questioned our ability to bridge the gap between material remains and the meanings ascribed to them by archaeologists (Courbin, 1988). In popular accounts of archaeology, there is the joke about every new find being ascribed to some 'ritual' purpose (Hodder, 1995). Although many archaeologists would argue that their subject is not a science (and, hence, all claims are subject to uncertainty about truth and subjectivity of interpretation and analysis), modern archaeology is based on a number of scientific methods and assumptions (Trigger, 2006). Some of these key assumptions are: the world is uniform, and changes occur at regular intervals (no randomness); artefacts that are datable, such as coins, can be used to confirm the earliest possible date of use of a site, but not the latest (that is, a coin minted in 380 CE cannot have been used before that date but may have been used for many years after that date); small changes to the sophistication of artefacts can be seen as evolutions in culture over time; and as layers of debris/habitation occur over time, the newest layers are at the top and the oldest are at the bottom. This allows a science of comparing and dating to work. Another more debatable assumption is that sudden discontinuities in the archaeological record (such as the sudden absence of coins) suggest a quick change in culture (Hodder, 1995).

Archaeology can confirm, disprove and question history, and it is especially important where primary sources do not exist (pre-history) or may be biased or partial. Historians use archaeology to write more sensitive, cautious histories, based on the evidence of finds. But if history is used to explain or analyse archaeology, there is a danger that by then using archaeology as 'evidence', there is a circular argument (Courbin, 1988; Trigger, 2006). What needs to happen is a synthesis of evidence and critical interpretation (see Afterword section below).

In this chapter, I have tried to bridge the gap of prehistory by identifying continuities and breaks between deep history, our early hominid natures, our emergence as modern *Homo sapiens sapiens*, and the pre-history of oral cultures. I have argued that there is a continuity of leisure experiences in all these moments in evolution, time and space, evinced in the material finds, from the cave-paintings of Lascaux to the ball courts of the Aztecs. It may be that not every interpretation of the material record is correct. I have relied on secondary sources, the interpretations that archaeologists have made. Some of them could be wrong. Some of them will be subjective and biased towards thinking in certain ways about the things they find. But there is an overwhelming amount of material data in the record that can only be interpreted as having a leisure function (Salamone, 2000). To find some communicative solace in leisure, away from the demands of work and food-gathering, is what marks us as human: the ability to think and reflect on the leisure activity, to make meaning out of the activity, to find the abstract of the soul in the trivia of the practice. However, in all times, and in all places and all cultures, there is evidence of another trend in human leisure lives: an instrumentality that bends such leisure to its will, whether that will is the religious and ritual demands of the temples, the demands of insane despots, the pride of the tribe, the insidious temptation of capitalism or the invisible Gender Order of hegemonic masculinity (Connell, 1995).

Afterword: the evidence for Julius Caesar

If all we had about Caesar was his own works, then that would be enough to satisfy most critics. He published his account of his Gallic Wars, and his own exploits in the Roman Civil War. These were widely read and referenced by a huge range of independent, contemporary sources and have come down to us intact through the process of reproduction. We also know Caesar from his own letters. Some of these survive with the Letters of Cicero, which were published after Cicero's death by his followers. In Cicero's letters, we hear Cicero and his correspondents mention Caesar many times. Caesar is also written to by Cicero, and he sends letters to Cicero in return. (Famously, Cicero asks after his relative, who is with Caesar in Gaul.) Caesar's letters are very short and to the point, and they clearly indicate that he has no time for the bluster of Cicero.

We also know Caesar from contemporary work published about him, especially in the period immediately after his death. Cicero's letters to his murderers are preserved, as are their replies. We also have the *Res Gestae* of Augustus, Octavian, Caesar's adopted son and heir, which discusses in great detail the circumstances of Caesar's death, and his deification. Augustus styled himself the Son of God, the Son of Caesar.

In the years following his death, Plutarch wrote his *Lives*, which included a biography of Caesar. This has survived, along with the later work of Suetonius on the Twelve Caesars. These are rich sources for historians – but, of course, problems of objectivity, distance and corruption make them as questionable as any other text surviving from the first and second centuries CE that attempts to tell us of earlier events.

If the writing was not enough to satisfy us, then we can achieve consilience about the existence of Caesar and those around him through other independent sources (consilience was first used by Victorian natural philosopher Wiliam Whewell to describe how we use the gathering of different pieces of evidence to build up a picture of the truth). These include coinage with descriptions of Caesar's titles alongside his name (there are hundreds of surviving coins, many of them found *in situ* by on archaeological sites), inscriptions across the early empire mentioning Caesar or commemorating him (or worshipping him), the inscriptions of *Res Gestae* of Augustus, but the clinching pieces of evidence are the Fasti Consularis and the Arval Brotherhood. The Roman Republic, at some point in its early history, started to keep records of who were the two Consuls of the year, and also (through the Arval Brotherhood) what happened in a given year. These two separate almanacs – the calendar of the Arval Brotherhood and the Fasti – record contemporary names and allow us to link names in one with the other. The Fasti survive intact from the later Republic and early empire, and it is in this list that we can find Gaius Julius Caesar, declared Consul five times from 59 to 45 BCE, the year of his assassination.

Of course, you could claim that all this is faked, or that none of it can be dated to the time of Caesar, or that the nails in the door of the Arval Temple that recorded the passing of every year were just put there by a trickster deity. But this combination of historical and archaeological evidence is enough to be compelling, and no serious historian doubts the existence of Caesar: we may as well doubt the existence of the Moon.

4
Leisure in Classical History

This chapter concentrates on the Mediterranean world of Classical Greece and Rome, where there is a wide range of primary and secondary sources to guide us from Cicero and Suetonius to contemporary historians such as Balsdon (2004) and Golden (1998). In the first section of this chapter, the role of games in Hellenic and Hellenistic culture will be contextualized in wider debates around civilization and masculinity. I will use the games to discuss the relationship between the sacred and the political, and I will argue that the moral and ontological ideal of the leisured citizen, drawn from the work of Plato, does not represent the complex consumption of these games. In the second section of this chapter, I will take the story of games to Rome, and return to the question raised in Chapter 1 of how similar the games of the Roman Empire were to the spectator sports of late modernity. I will argue that Roman games cannot be understood out of the context of their meaning for spectators and their purpose for the state in the Early Roman Empire. In the third and final section of this chapter, I will take this specific discussion of the meaning and purpose of games to explore the leisure lives of the Romans, as expressed through the condemnations of elite writers such as Marcellinus Ammianus and the evidence of archaeology. I will argue that leisure was an essential quality of *Romanitas*, what it meant to be a true Roman man, but that this sense of *Romanitas* was, paradoxically, based on an appreciation and appropriation of Greek culture and leisure forms, and a rejection of the Roman Empire's popular culture. In arguing for communicative action in shaping leisure choices, I will use the work of Pausanias (his guide to the sites of Greece) and the example of the Emperor Hadrian's Egyptian adventure to argue that increased wealth and mobility allowed a form of heritage tourism to establish itself.

Hellenes

Culturally speaking, the world of the Ancient Greeks (the Hellenes as they are called in the classical period), from the sixth century BCE through to the first centuries of the Common Era, was unlike anything that had gone before it. It is true that the Hellenes inherited myths, attitudes, gods, political systems, traditions and superstitions from their immediate antecedents in the eastern Mediterranean, such as the Persians, the Egyptians and the Lycians (Fox, 2005). There is no doubt also that the Hellenes' mercantile sense of adventure and colonization, and their use of the written word through the alphabet, was adopted from the Phoenicians, a semitic people who lived mainly in what is now the Middle East and North Africa (Ibid.). However, what made the Hellenes different from their predecessors and their neighbours were their pursuit of knowledge, their commitment to learning, and their attempt to define the good life of the true Hellene man in a way that transcended their local culture and moment in time to become something that was seen as universally civilized for over 2000 years.

This impact of Hellenism, historically, was the result of their promotion of writing and literacy as a vehicle of philosophical discovery and creative invention. Previous literate cultures had used writing as an imperial tool: the spreading of propaganda, as witnessed in the *stelae* of Persia or Egypt; the lists of goods needed by bureaucrats in the centralized planning system of Babylon; or the holy scripture of Israel, which marked out its people as chosen by God, and its priests as protectors of that scripture (Houston, 2008). The Hellenes, uniquely for the time, saw literacy as a tool for entertaining – through formal readings of books, poems and the performance of plays – and educating people in the right way to think (Braund, 1994; Rhodes, 2003; Fox, 2005). Crucially, the Hellenes rejected the hegemony of imperial states, preferring to experiment with the rule of oligarchs, or the rule of democracy (Fox, 2005). There were Ancient Greek dictators, of course, and cruel despots and kings. It also must be acknowledged that the Hellene civilization was based on continual warfare and plundering, the keeping of slaves with no rights, and the sequestering of women in domestic spaces. But the Hellenes – that small class of free men, and especially those free men with wealth – saw themselves as the bearers of a tradition of freedom, free speech and free thought (Braund, 1994).

The Hellenic civilization, through the empire-building of Alexander, grew into the Hellenistic civilization of the East – and whatever the Eastern influences on that civilization, the ideal Hellene man, expressed in the cultivation of free thought, of manly sports and of leisured freedom, remained the marker of what it meant to be a true Greek (Golden, 1998; Papakonstantinou, 2002, 2003; Fox, 2005). To be a true Hellene man was to demonstrate, publicly, one's commitment to that tradition, to its civilization, to its Hellenic and Hellenstic masculinity. That meant, in practice, engaging in political life, standing for public office, funding civic buildings such as temples and gymnasia, listening to other people's ideas in public spaces. As Habermas describes the ideal Greek city-state:

> In the fully developed Greek city-state the sphere of the polis, which was common (koine) to the free citizens, was strictly separated from the sphere of the oikos; in the sphere of the oikos, each individual is in his own realm (idia). The public life, bios politikos, went on in the market place (agora), but of course this did not mean that it occurred necessarily only in this specific locale. The public sphere was constituted in discussion (lexis), which could also assume the forms of consultation and of sitting in the court of law, as well as in common action (praxis), be it the waging of war or competition in athletic games... In the competition among equals the best excelled and gained their essence – the immortality of fame. (Habermas, 1989[1962], pp. 3–4)

The golden age of fifty-century BCE Athens, home to Socrates, the founder of systematic reasoning through logical argument, and Pericles, the author of the first democratic constitution, is in the mind of Habermas as he discusses the ideal Greek city-state. This city-state is legitimized through a sleight of hand: its (wealthy) male citizens are enfranchized and given the leisure time to discuss matters of public importance through the mechanics of slavery and domestic suppression of women. What was common to the free men of each city-state was also common among those city-states: the geographical and cultural expansion of the Hellenic world also spread its solutions to debate and competition. For Habermas, those solutions included the seasonal warfare of the elite classes but also the regular cycle of sacred games.

It has become a tradition of modern Olympians to express the belief that the Ancient Greeks saw in the sacred games an amateur, manly virtue of physicality that was equally important to them as their leisured learning. This story is one that is common in academic histories and sociologies of sport, too: that the round of games associated with the worship and veneration of gods and sacred places (of which the Olympics was and is the most well-known) became an expression of perfect Hellene masculinity (Golden, 1998; Coakley, 2003; Fox, 2005; Papakonstantinou, 2010). Of course, there is an evident truth in this story. These games were a way of bringing disparate, wealthy, free Greek men together to celebrate male athleticism and prowess (Papakonstantinou, 2003). There was a clear relationship between the sacred and the political in these games, honouring the gods and honouring the Greeks, helping cities boost their prestige and independence at times when empire-builders such as the Persians, Alexander, the Spartans or the Romans were on the horizon.

However, as Young (2005) points out, there was an ambivalent relationship between the participants in, and the supporters of, the games, and the intellectual and cultural elite from Plato into Hellenistic times and the Roman period. The athletes who achieved victories in the names of various gods were not amateurs dabbling at sports like Victorian muscular Christians: they were in all respects fully professional and instrumental in their approach to preparation, including regimens of diet and strategies for performance-enhancement (Young, 2005). The winners were celebrities of their time, adored by their home cities, honoured with gifts and hard cash. This was a reality of commerce and competition that many of the leisured elite – including Plato, Aristotle, Xenophanes and Isocrates – looked upon with patrician contempt (Young, 2005). The moral and ontological ideal of the communicative, leisured citizen, drawn from the work of Plato, does not represent the complex, instrumental consumption of these games. As Young (2005, p. 35) puts it:

> In antiquity, there was nothing at all about the mind associated with athletics of the Olympic Games. The notion that such big bruisers such as the ancient wrestler Milo were somehow akin to our fictitious scholar-athletes is just another Olympic myth; it is unclaimed baggage left behind my the myth of Greek amateur athletics.

Games in Rome

The Greek model of the city-state survived, according to Habermas, as a model of the ideal state of affairs between free citizens through the Hellenistic age and into the Roman world. The Roman Empire, though de facto clearly an autocracy, and often an arbitrary one at that, was itself legitimized by myths about the balance between the *Princeps*, the Senate and the People of Rome. Even at its most powerful under the 'Five Good Emperors' of Edward Gibbon, the propaganda of the Roman

Empire continued to claim decisions and conquests were made in the name of the latter two groups (Birley, 1993; Gibbon, 2005[1776–1788]).

The Roman Republic's genesis is the subject of a number of myths and narratives, which have come to us as history in the work of Livy. The Romans themselves dated the founding of their city to the eighth century BCE. They believed the city-state was originally a monarchy and that this monarchy was eventually overturned by a revolution of the patrician classes, which led to the quasi-democracy of the Republic. How much of the famous mythology of the founding of Rome and the Republic is true, and how much false, is debatable (Bringmann, 2007). What is clear is that, under the Republic, Rome established laws and a constitution based on a balance of rights and power between magistrates representing the senatorial elite (the consuls, two of whom were appointed to govern for 12 months), and those representing the free men of the citizenship (the tribunes). Under this system, the Republic entered written history as it expanded and started wars with other Mediterranean city-states. Rome's expansion was founded on the belief that *Romanitas*, what it meant to be a true, free, Roman man, a citizen of Rome and a member of its ruling classes, could be learnt through disciplined warfare and civic voluntarism (McDonnell, 2009). As the Republic's empire expanded, this narrow definition of *Romanitas* spread through Italy, then further through the establishment of Roman provinces and colonies from Spain through to modern-day Turkey (Scullard, 1982; Bringmann, 2007).

With Roman culture firmly established in Roman colonies, it was inevitable that the populations conquered by the Roman Republic would be influenced by a wider cultural form of *Romanitas*, even if the precise nature of such an influence (enculturation, hegemonic or Foucauldian interaction) is of great debate by modern-day historians (Mattingly, 2007). What matters to our history of leisure is the gradual spread of Roman leisure assumptions and practices, as evidenced in primary sources and, more clearly, in archaeological remains across Europe, North Africa and the Middle East. Across the Roman world, free men were able to exercise their *Romanitas* by watching games, listening to book-readings, buying take-away snacks, dining in relative opulence at home, and visiting taverns and brothels at night (Balsdon, 2004). Women, the poor and slaves were denied some access to public spaces but were still able to watch games, albeit from cheaper seats (Fagan, 2011). Bathing in specially-designed, publicly-built bath-houses became a commonplace, even in areas where urbanization was limited (Yugul, 2009). Bath-houses were available for men and women, rich and poor, though with strict rules of gender and class segregation (Balsdon, 2004; McDonnell, 2009). At the old Roman town of Ribchester in Lancashire, England, far from the Mediterranean, is a bath-house associated with a *mansio*, a wayside inn. If one passes through the museum one can see backgammon-like counters for board games, and pieces of pottery that were once part of containers for wine or beer. Nearer to Rome, the buried town of Pompeii, rediscovered under the volcanic ash of Vesuvius, demonstrates the importance of taverns, gladiatorial games and brothels in the leisure lives of Roman men in the time of Emperor Titus: leisure there is known to us from the buildings uncovered, the artefacts of everyday life, and the graffiti written in dozens of places promoting the value of foodstuffs, gladiators and prostitutes (Balsdon, 2004; Berry, 2007).

By the time Livy was writing his history, the Roman Republic, already de facto an empire, with colonies and provinces throughout the Mediterranean, was becoming an empire de juro. In the first century BCE, Roman politics had become dominated by factionalism and the power of generals. In public, these generals displayed all the characteristics of Roman virtue: they spent money on theatres and baths, and supported artists and writers through networks of formal patronage. They provided games when they became consuls, funding increasingly lavish entertainments that included chariot-racing and gladiatorial contests. These latter had become an integral part of Roman leisure, and gladiators were becoming celebrities. As Suetonius writes of Julius Caesar, in The Lives of the Caesars, written in the early second century (Julius Caesar 26, in Suetonius, 2003, p. 13), in his bid to be elected consul for the second time:

> He announced a gladiatorial show and a public banquet in memory of his daughter Julia... He also issued an order that any well-known gladiator who failed to win the approval of the Circus should be forcibly rescued from execution and kept alive. New gladiators were also trained, not by the usual professionals in the schools, but in private houses by Roman knights and even senators... Letters of his survive, begging these trainers to give their pupils individual instruction in the art of fighting.

The gladiator is the most famous icon of Roman leisure and sport, and probably the first person anyone thinks about when Roman games are mentioned. In some respects, we can see clear parallels between the gladiatorial games and our professional sport: athletes as celebrities

(Dunkle, 2008; Papakonstantinou, 2010); the spectacle of the cheering crowd (Fagan, 2011); the professional training (Dunkle, 2008); the gambling on outcomes; and the fanaticism of the die-hard followers versus the casual dalliance of the elite (Fagan, 2011). And just as some writers today sneeringly dismiss professional sport as something for the rabble to enjoy, so in the first century the Roman senator and orator Seneca (*Epistles 7,* found on-line at http://www.fordham.edu/halsall/ancient/seneca-letters7.html) could dismiss the games:

> I turned in to the games one mid-day hoping for a little wit and humor there. I was bitterly disappointed. It was really mere butchery. The morning's show was merciful compared to it. Then men were thrown to lions and to bears: but at midday to the audience. There was no escape for them. The slayer was kept fighting until he could be slain. 'Kill him! Flog him! Burn him alive' was the cry: 'Why is he such a coward? Why won't he rush on the steel? Why does he fall so meekly? Why won't he die willingly?' Unhappy that I am, how have I deserved that I must look on such a scene as this? Do not, my Lucilius, attend the games, I pray you. Either you will be corrupted by the multitude, or, if you show disgust, be hated by them. So stay away.

The morality of the gladiatorial combat is, of course, not just strange to first-century senator-philosophers: it is very strange to those of us in the twenty-first century, too. For the Romans, killing captured enemies in the games was a perfectly logical response to the problem of what to do with prisoners: the slaughter also reminded the Romans of the power of the republic and the empire, and the weakness of its barbarian enemies. From the slaughter of criminals to fights between condemned men, then fights between the best prisoners, trained in the art of combat, then fights between trained gladiators, fighting for money, adoration and freedom, the steps to the gladiatorial games are obvious (Dunkle, 2008; Fagan, 2011). In a society where people were at the mercy of *pater familias* in the private sphere, and the Senate and People of Rome in the public sphere, the utilitarianism of such slaughter would be evident. The gladiatorial games served to enhance the power and status of the magistrates, the state religions of Rome, and the cult of the emperor. They demonstrated to the Romans the superiority of their way of life, their civilization, over the Greeks and the Gauls and every other culture they had conquered. While responses to the gladiatorial games ranged from licence to loathing, all Romans saw the games as something natural to their leisure lives – and the state saw the value of giving these spectacles to a populace susceptible to riot and a military prone to rebellion and overturning one emperor for another.

In Chapter 1, I raised the question of how similar the games of the Roman Empire were to the spectator sports of late modernity. Roman games cannot be understood out of the context of their meaning for spectators and their purpose for the state in the Early Roman Empire. We do not watch our sports stars killing each other – though we do cheer on our football and rugby players as they push and tackle, and we do pay our money to watch boxers attempting to incapacitate each other (Pringle and Markula, 2005). In some sense, then, we are not very far from the killing arenas of the gladiators, and certainly some dystopian films such as *Rollerball* (1975, directed by Norman Jewison), *Death Race 2000* (1975, directed by Paul Bartel) and *The Running Man* (1987, directed by Paul Michael Glaser) have portrayed near futures where killing in the name of spectacle and profits is again part of our sports entertainment. For the emperors and the state, then, the games had an instrumental role in maintaining their hegemony. For the participants, the games offered a chance to become free or to create wealth. For the spectators, the games were about proving one's *Romanitas*, one's masculinity or status in the rigid hierarchy of gender, class and caste. For the merchants, the games were a source of lucrative income – there is archaeological evidence that vendors sold and advertised their products at these games (Balsdon, 2004). What we can see is a distinct and unique Roman way of consuming these games, which can be analysed as a combination of instrumental control and enterprise, and a communicative search for identity.

Roman virtues

In *The Meaning and Purpose of Leisure* (Spracklen, 2009, p. 20), in discussing the idea that leisure choices and practices are constrained by structures, I wrote:

> The hegemonic status of power at the end of modernity suggests that leisure, in its commodified state, is a way in which the ruling classes keep the working-classes ignorant of their oppression (Carrington and MacDonald, 2008). For example, while earlier theorists such as Veblen criticised sport for being savage, Theodor Adorno and others in the Frankfurt School drew parallels with religion as described by Marx, the oft cited 'opium of the masses', as being a vehicle for the suppression of the masses by totalitarian states... This view of leisure

as the diversion of the masses is, of course, an old one. Juvenal, the Roman satirist, wrote that the people of Rome in his day were happy to be given dole (bread) and entertainment (circuses) as diversions from engaging in political debate:

... iam pridem, ex quo suffragia nulli
uendimus, effudit curas; nam qui dabat olim
imperium, fasces, legiones, omnia, nunc se
continet atque duas tantum res anxius optat,
panem et circenses. ...

(Juvenal, Satire 10.77–81, in Braund, 2004)

Juvenal spoke for the senatorial classes at the time of the Principate, the golden age of the Roman Empire, mocking the masses and the new men, those of the equestrian and mercantile classes who cared more for material goods and a sniff of power than civic service and manly virtue. Juvenal was strongly critical of the increasingly despotic and arbitrary nature of the Principate, held as it was by soldiers like the Emperor Trajan. The dole – the giving of bread to the poor – was an old freedom of the citizens of Rome and a way for senators and emperors alike to demonstrate their wealth and benevolence. Circuses – athletic games, chariot-racing, theatrical performances – were expected when magistrates were appointed, funded out of their private capital.

In the second half of the fourth century, in the reign of the Christianizing Emperor Theodosius, an ex-soldier from the Greek-speaking east of the Roman Empire wrote an account of the history of the empire. This soldier, Ammianus Marcellinus, was writing in a classical historiographical tradition (Barnes, 1998). Although Greek in upbringing, he wrote in Latin. The history he wrote was in the tradition of earlier Roman historians such as Tacitus and was written as an attempt to defend Rome's imperial and pagan past at a time when Christianity was gaining ground as the empire's official religion. This history now exists only for the period 354–378 CE, and the rest had been lost. The books that survive, however, are based in many instances on Ammianus' own experience as an officer in the Roman Army, in the service of the mercurial Emperor Julian, derided in Christian historiography as the Apostate (Bowerstock, 1997). Julian was the last in the line of the Constantinian Dynasty. Unlike Constantine and his sons, however, Julian abandoned Christianity for a neo-Platonic paganism inspired by his schooling in Athens and his respect for the old ways of the empire. On being declared emperor by his troops in 361 CE, Julian was following in a centuries-old Roman tradition of emperors taking power by force. He avoided a war with his relative the Emperor Constantius when Constantius died. In power, Julian promoted paganism and revoked laws that favoured Christianity. For the empire's pagans, predominantly in the traditional decurion classes, Julian offered the chance of restoring the unity and strength of the empire. Julian's reign, however, was short: driven by his own overconfidence in his gods, Julian marched on Persia in a quixotic search for victory and was killed in his retreat from the land that had claimed the lives of many other Roman generals.

Ammianus' history is partly a paean to Julian, but also a rhetorical rage at Julian's failures and the incompetence of the emperors who followed him. These restored Christianity and encouraged corruption, indolence and a suspicion of learning and the serious leisure lives of the Roman elite (Friell and Williams, 1998). Ammianus retired from the army and went to live in Rome, where he wrote his history to please the few Romans who were still living lives of *otium*, learned and morally fit leisure, where books were read (or listened to at public and private readings). His opinion of the rest of Rome's citizens was low and has survived the part of the history that has come down to us. In Book 14.6 (in Ammianus Marcellinus, 2004, p. 48), he digresses on the ruling classes of Rome:

> In this state of things, the few houses which once had the reputation of being centres of serious culture are now given over to the trivial pursuits of passive idleness, and echo with the sound of singing accompanied by wind instruments or the twanging of strings. Men put themselves to school to the singer rather than the philosopher, to the theatrical producer rather than the teacher of oratory. The libraries are like tombs, permanently shut; men manufacture water-organs and lutes the size of carriages and flutes and heavy properties for theatrical performances.

On the lower classes, again in Book 14.6, he is equally scornful of what they do with their time (ibid., pp. 49–50):

> ...some spend the night in bars, others shelter under the awnings of theatres... They hold quarrelsome gambling sessions, at which they make ugly noises by breathing loudly through their nose; or else – and this is their prime passion – they wear themselves out from dawn to dusk, wet or fine, in detailed discussion of the merits and demerits of horses and their drivers. It is most extraordinary to see a horde of people hanging in burning excitement on the outcome of a chariot race.

Ammianus despised the immorality of the leisure activities of the Romans of his day. The ruling classes were vain and lecherous and preferred dancing, bathing and fashion to philosophy or improvement of any kind (Ammianus, 28.4). The lower classes, the proles, did nothing of any value, either, and were slaves to drink, vice and gambling. Ammianus was a believer in the traditional Roman values of virtue: of piety to the gods, respect for learning, and celebration of martial courage. In this sense, he was following the work of earlier critics of Roman vices, such as Juvenal, Seneca and Cicero, who all mocked trends in Roman culture that promoted dancing, gambling, drinking and sex at the expense of learning and military training. The belief in a lost ideal Roman, manly virtue was prevalent in the empire. Livy's history of the Roman Republic (Livy, 2005) mythologizes this virtue in the heroic tales of men bringing glory to Rome through their endeavours: whether on the field of battle, winning glory against impossible odds (e.g., Horatius on the bridge, or the senators who stayed in their homes, proudly sitting down to die, as the Gauls sacked the city) or in the realm of politics (Cincinnatus, who retired to his farm after saving the Republic).

In private, however, the conduct of the elite in the Principate was similar to that of the upper classes of the fourth-century CE, satirized by Ammianus. Cicero's famous *Philippics* against the conduct of Marc Antony portray that important member of the second triumvirate as a feckless hyper-sexualized drunk (Cicero, 2002). In *De vita Caesarum*, Suetonius is damning of those emperors in the first century CE who failed to be virtuous men in their leisure lives. Augustus, the first man to be declared Princeps, is characterized by Suetonius as a womanizer, but in all other respects a man of simple, unadorned and frugal tastes. Many other writers were equally generous to Augustus (Cassius Dio, 2005), and the emperor's own attempts to bring about a restitution of Republican morals and tastes are well known (Southern, 2001). However, beyond Augustus the line of emperors is described in predictably damning prose. On Tiberius, Augustus's successor, Suetonius writes (Tiberius 42–43, in Suetonis, 2003, p. 131):

> Even as a young officer he was such a hard drinker that his name, Tiberus Claudius Nero, was displaced by the name 'Biberius Caldius Mero' – meaning: Drinker of hot wine with no hot water added. When already Emperor and busily engaged in the reform of public morals, he spent two whole days and the intervening night in an orgy of food and drink... On retiring to Caprae he made himself a private sporting-house, where sexual extravagances were practised

> for his secret pleasure. Bevies of girls and young men, whom he had collected from all over the Empire as adepts in unnatural practices... would copulate before him in groups of three, to excite his waning passions.

In discussing the Emperor Nero, Suetonius provides an insight into the games and theatre that delineated (with the baths) the public leisure spaces of the early Roman Empire. Suetonius described how the young Nero was caught at school talking about a charioteer of the Greens (Nero 22, in Ibid., p. 224). In the same chapter, he says Nero played with ivory models of chariots, and was a serious follower of the races before he insisted in taking part himself. He also took part in musical contests, and 'broke tradition at Olympia by introducing a musical competition into the athletic games' (Nero 23, in Ibid., p. 225). In all these public spectacles, the citizens of Rome (or the city where Nero was performing) attended and cheered on their emperor at all the right moments. Nero, of course, insisted on winning all the prizes himself, whether for music, drama or racing. His self-absorption is not just a figment of Suetonius' imagination: we have other sources that are equally critical of Nero (Griffin, 1987), and it is no surprise that the emperor's immoral leisure practices were reflected in his immoral and arbitrary use of executions to try to keep in power. When some of his generals rose up against him, he fled the capital and, in a panic, committed suicide.

Those who replaced him were not much better, at first. The Emperor Vitellius is described by Suetonius as a glutton, who 'banqueted three and often four times a day' (Vitellius 13, in Suetonius, 2003, p. 276); Vespasian and Titus are better men; but Vespasian's son Domitian is as deprived as Tiberius, and as wilfully violent as Nero or Caligula. Suetonius says (Domitian 19–22, in Ibid., pp. 317–18):

> Domitian hated to exert himself. While in Rome he hardly ever went for a walk... His chief relaxation, at all hours, even on working days and in the mornings, was to throw dice. He used to bathe before noon, and then eat such an enormous lunch... Domitian was extremely lustful, and called his sexual activities 'bed-wrestling', as though it were a sport.

Suetonius, of course, was writing for a senatorial audience in an age when the empire was governed by good men: Hadrian, then Antoninus Pius. His colourful biographies are meant to titillate: they are the equivalent of our scandal-sheet tabloids, or celebrity biographies (Mellor, 1998).

Suetonius relishes the detail of sexual depravity and never misses a chance to rehearse the virtues that he believes make a good emperor, those he thought were the qualities of Roman manliness defended by Livy and Cicero (McDonnell, 2009). But it is clear that his book was believed and read as truth by the Romans of the second century CE. The depraved leisure lives were believed because people could remember life under Domitian, or they could see similar louche habits in their current Roman society.

What both Suetonius and Ammianus show us is the continuity of many leisure practices in the Roman Empire, from the Principate of the first century through to the Christianized fourth century (Ward, 1992). Romans in the reign of Emperor Valentinian behaved in their leisure in ways very similar to those of the reign of Emperor Nero: baths, games, brothels, taverns and theatres were still frequented; free men of wealth still defined their *Romanitas* through patronage and conspicuous consumption (McDonnell, 2009); women had limited, domestic spaces, apart from a few women from noble families with the disposable income to spend in a similar frivolous manner as the wealthy men (D'Ambra, 2006); the lower classes had limited time and space for leisure, especially in rural settings (Balsdon, 2004). As I will discuss in the next chapter, the growing influence and dominance of Christianity on Roman culture – in particular, Christian morals and ethics – would significantly change the Roman leisure sphere, and the meaning and purpose of leisure. What Christianity's rise demonstrated, though, was the continuing influence of Hellenism on Roman minds.

Tourism

In *The Meaning and Purpose of Leisure* (Spracklen, 2009, pp. 52–53), I compared the development of a hegemonic Romanizing culture with modern arguments about the globalization of Western popular cultural forms:

> In his monumental work *The History of the Decline and Fall of the Roman Empire*, Edward Gibbon sketches out the reasons for the success of Imperial Rome in the first and second centuries CE, so that he can then show better the way in which the empire declined in power (Gibbon, 1776–1788:2005). One of the key ways in which the Roman Empire accrued power, according to Gibbon, was the spread of *Romanitas*: the essence of Roman cultural norms and values, its civilization. In every corner of the Mediterranean world, people adopted Roman fashions, Roman games, Roman literature, Roman

food and Roman political systems – whether they were conquered by Rome's soldiers or just living in parts of the world in Rome's sphere of influence. Rome, of course, stole much of its cultural heritage from the places conquered by the republic and the empire. The Hellenistic world, in particular, was a strong influence on the Romans, and preserved a sense of Greekness throughout the centuries of Roman hegemony, to the point where the only part of the Roman Empire surviving was the Greek-speaking empire of Byzantium in the East (which survived until its destruction by the Ottoman Empire in 1453). But Byzantium proves the power of *Romanitas*: even hundreds of years after the fall of the empire in the West of the Mediterranean, the emperors of Byzantium maintained Roman political structures and dreamed of reconquering the West (Gregory, 2005). In the world known to contemporary Roman writers, *Romanitas* was something that transcended fixed notions of culture and belonging, something that imposed civilization and order in the boundaries of the empire and in the satellite kingdoms surrounding it (Potter, 2004). Although there were rules limiting political rights of Roman citizenship, inhabitants of the first and second century Roman Empire were Roman by their acceptance of and participation in Roman culture. By the third century, all free inhabitants of the Empire were Roman citizens, and the Roman Empire had already seen emperors born in Syria, Africa and Arabia. What united the Roman world was *Romanitas*. Beyond *Romanitas* was only a confusing world of barbarians who, to a greater or lesser degree, adopted various facets of the Roman world into their lives (Heather, 1998). To be Roman was to be part of a global civilization, sharing an understanding of Cicero, worshipping Jupiter and Mars, speaking and writing in Latin, watching games in *stadia* that have left so many standing remains in Europe, North Africa, Syria and Turkey. It is just a coincidence that the single arch of Roman architecture was the symbol of Rome's power and hegemony and globalization, in the same way that the double arches are a symbol of America's cultural colonization of the globe. It is because of this globalizing trend in Roman history that so much physical evidence of Rome survives for us to consider today. Romanization spread Roman literature, and literacy, so that when the empire collapsed in the West there was enough of a culture of literacy and learning to ensure the transmission of primary sources like the work of Tacitus.

Leisure was an essential quality of *Romanitas*, what it meant to be a true Roman man, but this sense of *Romanitas* was, paradoxically, based on an appreciation and appropriation of Greek culture and leisure forms, and a rejection of the Roman Empire's popular culture. Educated Romans were fascinated by the heritage, history and culture of the Greek world they conquered. At the time of the Republic, when the Romans consolidated their hold on the Greek East, Greek learning and practices became fashionable in Rome to an extent that defenders of Rome's traditional manly ways complained about the femininizing Greek influences (McDonnell, 2009). However, whatever patricians believed about the weakness of Greek culture, the wider population in Rome accepted a number of Greek forms into their lives. Greek historians such as Herodotus and Thucydides were taken as exemplars for Roman historians (Mellor, 1998); Greek philosophers, especially the Stoics, became highly influential in the way Romans thought about morals and virtues (Thorsteinsson, 2010); Roman poets and playwrights copied Grecian forms; and Roman engineers such as Vitruvius borrowed from the natural philosophy of Archimedes and Hero of Alexandria.

From Suetonius, we can read that the Emperor Nero was enamoured of Greek leisure and culture: Greek sports and games, music, arts and drama. While Nero was not necessarily a good role-model for Romans wishing to follow the trend, there is evidence that the late first century CE saw an increasing appreciation of the Greek world in the Roman senatorial classes, which in turn influenced wider Roman society when some upper-class satirists complained about the negative and populist influence of what they described as the un-manly Greek habits and interests (McDonnell, 2009). By the second century CE, Hellenism had become a common-place among Romans. The Emperor Hadrian spent much of his wealth rebuilding civic spaces in Athens and the Greek East, and supported the revival of pan-Hellenist political structures (Birley, 2000; Goldhill, 2001). As well as this cultural turn in policy, Hadrian decided to see the sites of the wider Hellenistic world himself, replacing the traditional imperial summer season of warfare on the borders with extended tours of Greece, the Middle East and Egypt. In the south of Egypt, a member of the imperial household inscribed her name on the whistling statue of Memnon, as was the tradition for travellers visiting it (Danziger and Purcell, 2006). These tours took the entire imperial household over vast distances, promoting the development of towns on the route, and a myriad of small businesses promoting sites of interest, souvenirs, local foods and transit stations (Birley, 2000). The Emperor Marcus Aurelius was not as ostentatious as Hadrian, but still lived a profoundly Greek life based on Stoic philosophy and ideals of leisured learning he himself had learned from his Greek tutors (Birley, 1993).

To meet the demand for knowledge about what Greek sites were of possible interest to travellers, and where the best places to stay might be found, early forms of what we know as guidebooks were published. Greek historians such as Herodotus had, of course, written about exotic cultures and foreign places, and these works could be used by a proto-tourist industry: many of the Greeks who followed the expansion of Alexander's empire into the Persian East, for example, were familiar with Herodotus said about the cultures there (Wallbank, 2010). Likewise, the Greeks were fascinated by Egypt, and the work of Herodotus shaped that fascination to an extent that Greeks in Alexandria wrote primers on Egyptian history (Kemp, 2005). The guidebooks of the second century CE emerged to meet a different audience – Roman, not necessarily Greek (though of course many Greeks were politically and culturally Romans) – and different need – being seen to be visiting, not necessarily understanding (Goldhill, 2001). The most famous that has come down to us is Pausanias' *Description of Greece,* written in the second century CE. Pausanias was a Greek, writing about his own culture, the places he knew and the stories he had heard, for a Roman audience. As Elsner writes:

> Pausanias chose to travel in and write about his own native land. He himself was aware that this was somewhat peculiar:
>
> The Greeks appear apt to regard with greater wonder foreign sights than sights at home. For whereas distinguished historians have described the Egyptian pyramids in the minutest detail, they have not made the briefest mention of the treasury of Minyas and the walls of Tiryns, though these are no less marvellous (ix.36.5)
>
> Greek writers preferred to turn their gaze upon the foreign than upon the self. The strangeness of Pausanias' enterprise lies in recording the monuments and rituals of his own society rather than those of other peoples. (Elsner, 1992, p. 7)

When it was first translated by modern scholars, it was doubted whether the book had been used at all for the purpose of educating and enlightening travellers: it was argued that there was no trace of its influence on other writers, hence no evidence of its use or impact (Alcock et al., 2001). However, more recently, scholars have recognized that Pausanias' book did have an impact on later writers, who were influenced by its style and structure, or who copied from it (Elsner, 1992; Snodgrass, 2003). Pausanias, like all travel writers, wanted to enlighten, but he also wanted to construct a particularly Hellene, non-Roman version of Greece in the minds of his readers (Pretzler, 2004). Furthermore, Beard and Henderson (2000),

Alcock et al. (2001) and Snodgrass (2003) argue separately that Pausanias' book was used in the purpose he intended it: travellers did take the book's information with them (either in hand or redacted into an abbreviated form or in memory) when they visited the Greek East. For there were many opportunities for such travel: officials and soldiers moved around the empire on the business of the state; merchants did the same in the search of profits; members of mystery cults and other religions, or potential members, went to holy sites such as Eleusis in Greece; and the wealthy had time and money to follow Hadrian's example. Even in the time of the Republic, the senatorial and equestrian elites had decamped to villas on the coast or in the mountains when the muggy summer in Rome became too difficult to bear. Sons of such families would often travel to be educated (McDonnell, 2009).

So we can see that the provision of opportunities to travel, the need as well as the desire, led to a demand for knowledge and information about parts of the empire unknown to the Romans who wanted to travel (or who wanted to know about the other parts of the world): Greece, of course, was the most obvious place of tourist desire, as it was the Greek world, and its culture and leisure, that was deemed authentically civilized. Pausanias, then, was the first writer to invent the tourist pilgrimage for authenticity (MacCannell, 1973). We can see a distinctly Habermasian communicative action in shaping leisure choices in the work of Pausanias, and the example of the Emperor Hadrian's Egyptian adventure – bearing in mind that the choice was only available to the wealthy, the free and the literate. Individual Romans were choosing to educate themselves about Greek history and culture. They were willing to invest time and money to travel to see places in the Greek world that fascinated them, whether religious sites, philosophical schools in Athens, venues of various games including those of Olympia, places associated with Alexander or Homeric heroes like Achilles or simply places with local folk or supernatural interest. The hegemonic power of the Roman Empire in the Mediterranean made travel safe and relatively simple. The cultural preference for Hellenism directed Romans to Greece, Greek culture and Greek leisure, rather than (say) the provinces of Gaul. Ultimately, the increased wealth and mobility allowed a form of heritage tourism to establish itself in the time of the five good emperors of the Principate.

5
Leisure, Islam and Byzantium

This chapter follows the Roman Empire into the Byzantine and Islamic cultures that succeeded it (Cameron, 1976, 1979). The first section of this chapter will begin with the Greens and Blues, first met in Chapter 1. A wider account of the role of leisure, games and elite and popular culture in the Byzantine Empire will contextualize the story of the circus factions. Procopius' *Secret History* (1981) will be used to present an account of the activities of the circus factions, and the leisure forms that were controlled by these factions. I will examine Christian writers and moralists of the early Byzantine period to explore the ways in which leisure was restricted and channelled. I will argue that in the increasingly autocratic eastern Roman Empire, theological debate became an outlet for communicative discourse and agency, to the extent that – especially in the sixth century – it became a form of leisure among the urban population of Constantinople. In the second section of this chapter, I will examine the restrictions and opportunities for leisure offered by Islam in the Early Caliphate Period, and the way in which leisure practices soon became a way of distinguishing among elites, the faithful masses and other People of the Book such as Christians and Jews. In the third section of this chapter, I will use the writing of Constantine Porphyrogenitus, Psellus and Anna Comnena to chart the re-emergence of Byzantine power and the ritualized, gendered leisure of the empire in the 900s and 1000s. In the fourth and final section, Byzantine leisure lives will be compared and contrasted with the leisure lives of the Ottomans, who succeeded to power in Constantinople in 1453. I will argue that both states routinely used leisure in similar instrumental ways, to ensure the continuity of political stability, communal identity and inequalities of power. At the same time, the urban nature of the centre of power ensured alternative, communicative expressions of leisure continued to be found in informal and commercial social exchanges.

Blues and greens

Byzantine leisure life was, in many respects, a gradual evolution of everyday, public and private leisure forms of the Roman world. In public, men gathered in the *fora* to talk, or met at taverns and brothels. The imperial family and other aristocrats supplied funds for games and races, and, in most cities of the empire, public baths. In private, women's leisure lives continued to be constrained by their low status, though this time the religious ethics that bound them to domestic duties came from Christianity, not the masculine virtues of Roman paganism. Children of the elite had access to hunting, book-learning and entertainers. The poor, including slaves, had little leisure other than the freedom to cheer their favourites in the circus, to gamble and to drink. This, of course, should not be surprising, for the Byzantines by self-definition were Romans, they lived in the Roman Empire, and they still considered themselves inheritors of Roman Republican cultural traditions (as well as Greek ones). It is modern scholars who call the later Roman Empire that consolidated around the eastern capital of Byzantium the Byzantine Empire. The Byzantine Empire continued Roman culture and political power through to the Renaissance: the capital fell to the Ottomans in 1453. The Byzantine Empire owed its origins to the Emperor Constantine the Great, who rebuilt Byzantium and renamed it (after himself) Constantinople. This became the capital of the Roman Empire while he still reigned, and afterward it retained its primacy as one of the Empire's imperial capitals. In Byzantium, Constantinople and his family built new churches, as befitting their conversion to Christianity – but they also built or supported the construction of baths, and a great hippodrome for chariot-racing and other spectacles. This hippodrome was built directly next to the imperial palace, with a covered passageway between the two for the Imperial court to move from one to the other.

Circus factions had been around in the Roman world ever since the final years of the Republic. Originally, there were four teams in the circus at Rome, identified by their colours: white, red, green and blue. The development of those teams over the next four hundred years is unclear, but some key patterns emerge. First of all, the teams became staffed by professional sportsmen, with trainers, stablemen and networks of support similar to those of the trades guilds and brotherhoods that dominated life in Rome in the first century CE. Second, as chariot-racing and other circus entertainments spread with *Romanitas* into the provinces of the empire, so did the circus teams. Third, the four colours became two, Blue and Green, as their wealth overshadowed the other two, which merged into the Blues and Greens or fell into disuse (Cameron, 1976). Finally, the two teams of Blues and Greens grew into factions, with supporters in the stands and on the streets, from the meanest tenement to the imperial palace. Those supporters were not, however, simply passive spectators, consuming wine and sweetmeats in the cheap seats. They were also effective street-gangs, private militias, theatre entertainers, singers, acrobats, prostitutes, zoo-keepers, managers, dealers in various black markets and carefully-orchestrated rioters.

By the fifth century CE, the circus factions had grown to dominate political struggles in the Byzantine Empire, throwing their muscle, money and contacts behind the different positions in political and religious arguments (Cameron, 1976, 1979). The Greens tended to support theological arguments in favour of monophysitism, for example, where Jesus was said to have one nature (Freeman, 2009). This led the Greens to back Emperor Anastasius, who was tolerant of (or in favour of, the truth is unclear – see reference) such anti-Chalcedonian (anti-Orthodox) views. The Blues backed the status quo of the Christianity and nature(s) of Jesus established by the Council of Chalcedon, and hence in the early sixth century CE they backed the emergence of the orthodox Justinian as the de facto ruler of the empire in the name of his uncle, the illiterate soldier Justin. When Justinian became emperor in 527, however, he had to try to manage the demands of both factions, to stop the Greens from being a focus of insurrection, or the Blues from becoming too powerful and corrupt (Evans, 2000).

Fortunately for our exploration of early sixth-century Byzantium leisure lives, we have an invaluable primary source from the period. The historian Procopius was close to the imperial court in Justinian's reign, closer still to a circle of writers and poets who encouraged him to write. He served as a secretary to one of Justinian's most famous generals, Belisarius, who recaptured the province of Africa for Justinian, and who invaded Italy and took Rome, bringing about a long Byzantine engagement in the former Roman provinces of the west (which had all fallen into the hands of barbarians or usurper rulers by the year 476 CE). Procopius wrote a number of histories for public consumption and accolade. But he also wrote a private, secret history of his times, where he was caustic and bitter about the emperor and many others, as well as Byzantine cultural and public life more generally. He was particularly scathing about the circus factions. In The Secret History, Procopius writes (2, in Procopius, 1981, pp. 71–72):

> The people have long been divided into two factions... [Emperor] Justinian attached himself to one of them, the Blues, to whom he had already given enthusiastic support, and so contrived to produce universal chaos... Needless to say, the Green partisans did not stay quiet either: they too pursued an uninterrupted career of crime, as far as they were permitted, though at every moment one or the other was paying the penalty... To begin with, the partisans changed the style of their hair to a quite novel fashion, having it cut differently from other Romans. They did not touch moustache or beard at all, but were anxious to let them grow as long as possible, like the Persians. But the hair on the front of the head they cut right back to the temples, allowing the growth behind to hang down to its full length in a disorderly mass, like the Massagetae... Then as regards dress, they all thought it necessary to be luxuriously turned out, donning attire too ostentatious for their particular station... The part of the tunic covering the arms was drawn in very tight at the wrists, while from there to the shoulders it spread out to an enormous width. Whenever they waved their arms as they shouted in the theatre or hippodrome and encouraged their favourites in the usual way, up in the air went this part of their tunics.

The Blues and the Greens controlled a large portion of everyday life across the empire, defining and providing cultural and leisure activities. This control was linked to their use of their own political powers and their use as pawns in the great struggles over theology. It might be argued that the Circus factions were not really interested in religious controversies – certainly, many of the faction followers may have been just in it for a drink and a fight, and many of the faction's customers probably just wanted to see bears, dancers and – yes – chariot-racing (Cameron, 1976). The link between the leisure activities and the theological dogma is never entirely moral: there is no evidence that Christian values stopped the Circus factions engaging in gambling, prostitution and petty crime (Haldon, 2008). However, there is evidence that successive emperors in the fifty and sixth centuries CE legislated against various forms of dark leisure, and the great complier of law Justinian was particularly moral in his rejection of sin (Evans, 2000; Watts, 2004), even if Procopius argued that, in private, Justinian was an evil and sinful man (Procopius, 1981). This legislation is the evidence for a public debate, informed by Christian ethics, which must have shaped and constrained, in some way, the activities of the factions (Brown, 1989). What the theological positioning of the factions reflected was such a public discourse about the meaning and purpose of Christianity – the Blues siding with the Imperial nature of the Orthodox; the Greens siding with the more democratic, populist and parochial theology of Monophysitism.

In the increasingly autocratic eastern Roman Empire, then, theological debate about the nature of Christ became an outlet for communicative discourse and agency, to the extent that – especially in the sixth century – it became a form of leisure among the urban population of Constantinople. This public desire to speak and debate about the meaning and purpose of Christianity was unique to Byzantium and its empire: in Western Europe, the Church of Rome imposed a dogma of belief and practice, which every Christian had to adhere to or consciously reject. Such rejection would lead to the charge of heresy. In Byzantium, on the other hand, it was possible, despite the violence of the state in imposing the beliefs of any given emperor on his subjects, to have and defend alternative positions: on the controversy over the nature of Christ, then, in the eighth century, on the controversy over icons (Barber, 2002). That possibility became a preference of many people's leisure time – not just educated courtiers, but also the artisans and traders who filled the streets of Constantinople. As Liudprand of Cremona (2007) observed, when visiting the city from the West, everyone he met had their own view about the nature of Christ and would freely offer their opinion to strangers in the hope of a friendly – but deadly serious – theological debate (see also Gregory, 2005). And when iconoclasm ruptured the Byzantine affection and use of icons, their use was preserved not by recusants burned at the stake, but by iconophiles in private houses, monasteries and even the imperial palace itself (Barber, 2002): iconoclasm failed in part because the Byzantines had got into the habit of using their public and private leisure spaces to think freely about religion. For iconophiles, iconoclasm felt too much like a rupture with traditions and cultural habits associated with the golden age of Byzantium; it also seemed to be a practice picked up by Christians in response to the fiercely iconoclastic religion that had emerged to the empire's south as the empire's most dangerous enemy: Islam.

Leisure and life in the early Caliphate period of Islam

In this section, I will examine the restrictions and opportunities for leisure offered by Islam in the Early Caliphate period, and the way in which leisure practices soon became a way of distinguishing among elites, the faithful masses and other People of the Book, such as Christians and Jews. Islam's origins have been the matter of much theological debate (Armstrong, 2001; Silverstein, 2010) and historical exegesis (Kennedy, 2004; Hourani, 2005; Donner, 2010). What started as a localized prophetic movement in parts of the Arabian Peninsula at the beginning of the seventh century CE spread rapidly to all the tribes of that peninsula, then beyond to the Byzantine and Persian empires. The energy and confidence of faith met two empires that had recently fought each other to exhaustion over Egypt, Palestine and Syria (Kaegi, 2003). In the Byzantine Empire, Monophysite and Nestorian Christians, as well as Jews, saw the benefit of a benevolent Muslim rule over the arbitrary punishments of Orthodox Constantinople.

After the Emperor Heraclius lost an army to the Muslims at Yarmuk in 636, the Byzantines lost Syria, then Jerusalem and the Palestine, then Egypt. By the end of the seventh century, Muslim armies had conquered the rest of North Africa (Silverstein, 2010), taking the Roman city of Carthage at the beginning of the eighth century CE. Muslim armies crossed into Spain, where they took advantage of suspicions and warfare among the Spanish to take over most of the province (Hourani, 2005). In the east, Muslim armies invaded and took over the entire Persian Empire in the seventh century CE (Kennedy, 2004). In the seventh, eighth and ninth centuries CE, Islam was a dominant force and culture in a large part of Europe and Asia, imposing or co-opting a hegemony of practice and belief on an *umma* that stretched thousands of miles from end to end (Lapidus, 2002; Hourani, 2005). Only the walls of Constantinople stood in the way of the Muslim Arab armies of this period, and slowly the Byzantine Empire lost the islands of Cyprus, Crete and Sicily to Muslim Arab conquerors (Kennedy, 2004; Gregory, 2005).

Yet, despite the story of conquest, there was a more subtle interaction among Islam, the Persian world and the cultures of Christian Byzantium and Rome. First of all, the Christian and Jewish populations conquered by invading Muslim armies were not routinely slaughtered. If they submitted to Muslim rule and paid taxes, these so-called People of the Book – of the Old Testament narratives that were shared by Islam, Judaism and Christianity – were left in peace (Kennedy, 2004; Donner, 2010). The relationship with the Zoroastrians of Persia was less clear, but essentially large populations of Persians were treated fairly similarly to the Christians and Jews of the Byzantine Empire (Silverstein, 2010). At first, the Early Caliphs of the Faithful instructed their soldiers not to mix with conquered populations, and not to encourage conversion

(Donner, 2010). Many people, however, did convert (Kennedy, 2004), and soon the artificial barriers between conqueror and conquered relaxed. Muslims were allowed to live in the towns and cities alongside the Greeks, Spanish, Africans, Egyptians, Syrians and Persians, and a syncretic culture emerged based on the practice of Islam and the popular cultural forms of Late Antiquity (Brown, 1989; Lapidus, 2002; Hourani, 2005). Baths, in particular, retained a prominent role in the leisure lives of Muslim towns in the Early Caliphate period (Lapidus, 2002). The division between public male spaces and private female spaces was adopted into Islam from Late Antiquity, too (Kennedy, 2004; Yugul, 2009). But some of the leisure activities of Late Antiquity – particularly drinking – were ruled out for Muslims, and the orthopraxy of the new faith also limited choices in other ways.

For Muslims, it is a matter of faith that Mohammed, a merchant from the Arabian town of Mecca, was divinely inspired by Allah to preach a unifying, monotheist message: the orthopraxy of Islam (Armstrong, 2001), written in the Qu'ran and reported in the numerous oral narratives that were ultimately codified in the *hadith* (Silverstein, 2010). Islam has a clear and simple message of faith in one God, and the word of God dictated to his prophet Mohammed, which proved incredibly popular among the Arabs of the seventh century CE (Hourani, 2005). Qu'ranic commitments to equality were unlike anything else seen in the region at the time: although women were still viewed as having fewer public freedoms than men, and slavery was not forbidden, it was still a radical message, and Mohammed attracted many followers in the mercantile classes denied power in the pre-Islam tribal hierarchies (Armstrong, 2001). Mohammed's teachings were initially condemned, and he had to flee Mecca with his followers when the polytheist rulers of the town feared his new religion would damage the pilgrim trade or *hajj* to the town's holy site of the Qa'ba (Donner, 2010). Mohammed returned to Mecca after some inconclusive battles, and more successful negotiations, saw the locals embrace Islam (Armstrong, 2001; Hourani, 2005). The *hajj* became one of the pillars of the Muslim faith, a pilgrimage expected to be undertaken by all Muslims at least once in their lives. Other pillars included two directly related to people's leisure lives: an obligation to pray five times a day, with set times associated with the rising and setting of the sun; and the commitment to consume only food and drink which is *halal*. Although the use of alcohol was the subject of differing stories and interpretations, and the words of the Qu'ran are not clear on this matter (Peters, 2007), it became a part of Muslim law, based on accepted hadith traditions, that alcohol was *haram* – banned.

After Mohammed's death in 632 and the rapid expansion of Islam into the Byzantine and Persian worlds, Muslims believed that Mohammed's life could be used as an example of how a perfect man lived (Ruthven, 1997; Armstrong, 2001). Stories were collected about how Mohammed ate, how he behaved in the street toward strangers, how much liberty he gave to his family and friends to question him, and other examples (Ruthven, 1997). He became an exemplar hero for Muslims – but unlike the heroes of the Romans and Greeks, he was humble, moderate, charitable and fair-minded (Armstrong, 2001; Donner, 2010). Men in particular judged their own commitment to Islam through the way they acted in public as private as closely as possible as Mohammed did. This meant, in leisure, avoiding anything that stirred the passions, and anything that manipulated the good will of others. Drinking, gambling, fighting, chariot-racing, sex outside the strict confines of domestic spaces – these were all things that were not morally correct. Music had an ambivalent position in the traditions, with some suggesting Mohammed liked listening to music, but others arguing that music was not acceptable (Peters, 2007). It was reported, however, that physical activity – sports and games – were important in the right leisure life of a free Muslim man (Farooq and Parker, 2009). As Stokes (1996, p. 23) writes:

> The Hadithic literature attests to the high significance attached to sporting pursuits, including wrestling. In one famous quotation, the Prophet claimed that the only leisure activities permitted to Muslim men were archery, horse-riding and making love to their wives (Uludag, 1976: 124–27)...This Hadith is often quoted in order to condemn most other leisure pursuits as haram...In a more positive vein, in other Hadithic injunctions, the Prophet comes over as something of a sports enthusiast. He is reputed to have been an excellent sprinter and rider. One Hadith (from Buhari) relates that the Prophet and his camel, Adba, were unbeaten, until a race in which they were defeated by a Bedouin rider.

The *hajj* pilgrimage played a key role in establishing travel routes across the Islamic world. As the religion took hold in various parts of the world, existing roads were upgraded, and new ones established, along with caravanserai or hostels at regular stages along the way. This was a policy from the Persians and the Romans, adopted and adapted by the merchants and pilgrims of Islam (Lapidus, 2002). These roads and way-stations were more than outposts of imperial authority: they became places of cultural and economic exchange, places where travellers could take sustenance, gossip and listen to stories (Lapidus, 2002; Kennedy, 2004). Merchants followed the soldiers and pilgrims, and roads, navigable rivers and coastlines became common ground for men of independent means, or men on pilgrimage. The road network connected urban locations where various mercantile guilds controlled a range of activities and industries for travellers; for pilgrims on the *hajj*, local emirs and sultans were responsible for purchasing the goods and services they needed to get through the territory on their way to and from Mecca (Hourani, 2005). In both instances, small-scale capitalist enterprises soon became sources of regular profits for the people running them, and a nascent tourism industry started to appear. The instrumentality of the *hajj*, and the instructions to assist pilgrims, made long-distance travel a relatively attractive activity. The communicative habit of travel made other sites, such as Jerusalem, pilgrim sites for Muslims seeking to extend their commitment to the *hajj* with a more personal piety. Pilgrimage also became something of a habit for Christians at the time of the rise of Islam: the belief that by making a long and arduous journey to a holy site one could be closer to heaven was a pernicious one, as we will see in the next chapter.

By the eighth and ninth centuries, the collections of *hadith* were being definitively established, and Islamic scholars were exploring and examining how the *hadith* and the Qu'ran could be used to interpret everyday life and morality. Some scholars argued for a jurisprudence based on an acknowledgement that the orthopraxy of the seventh-century Arab world was no longer completely useful and ethically correct for an *umma* that stretched across so many different cultures, which encompassed so many large cities filled with a host of people, religions and languages (Silverstein, 2010). A relaxation of the laws against casual sex and the use of alcohol was de facto something that emerged in the inner courts of the Caliphate itself (Kennedy, 2004), a trend that shocked more traditional scholars, but one that was mirrored in the habits of other urban elites (Lapidus, 2002). A cycle of relaxation of morals in leisure habits, followed by conservative reaction, followed by relaxation, appeared in the culture of the Caliphate: a similar cycle of central control and regional revolutions emerged in the political sphere, but it was not always clear – despite regular claims by conservative scholars – that moral lapses were the cause of Islam's setbacks. What is clear is that the Caliphate itself, by the tenth century, lost power to its own generals and sultans (Kennedy, 2004), who brought moral certainty to some parts of the empire, while the Caliphs themselves became shadow figures, hidden away in the pleasure rooms of their palaces.

The sultans who caged the Caliphs were Turkish in origin, and although they were Muslims, they brought into the Islamic world particular cultural practices that pre-dated their conversion. The Turks were traditionally nomads and brought with them a culture dominated by horses: horse-riding, trading horses as commodities, hunting and fighting using bows fired from the saddle, eating horsemeat and drinking fermented horse-milk. This love of archery and horse-riding was already present in the Islamic world, particularly in the former Persian provinces, but the Turks popularized hunting, archery, horse-races and other sports played out of the saddle (Stokes, 1996; Hourani, 2005). The Turks also brought to Islam a love of wrestling, which they may have adopted from the Byzantine Greeks (Stokes, 1996); this did not get taken up by Arabs in the Islamic world, but it did become popular in the eastern reaches of the *umma*, as well as in the Central Asian heartlands of the Turks.

In this time, Islamic morals over what is acceptable in a person's leisure life became more widely adhered to: men took over the public sphere more forcefully, women were relegated to private spaces, orthopraxy was followed in diet, health, sexual relations, physical activity and interactions with others (Ruthven, 1997; Lapidus, 2002; Silverstein, 2010). In public, men looked not only to the life of Mohammed for guidance, but also to the first four Caliphs, who became part of a social memory of Islamic virtues tied up with a narrative of righteousness and victory. Their stories and those of other early Islamic heroes set a tone for a noble, morally correct, fair and wise soldiery that – as we will see in the next chapter – played a crucial role in the construction of Western European notions of chivalry and aristocratic leisure.

Ritualized and gendered leisure in Byzantium

In the tenth century and into the first quarter of the eleventh, a succession of successful emperors and generals restored Byzantium to some of its former glory. A series of battles saw the empire's borders reach south towards the Holy Land, and north towards the Danube. Byzantine art and architecture reflected the confidence, wealth and power of Orthodoxy (Gregory, 2005). Byzantine traders were found throughout the Mediterranean and Middle East, Byzantine influence was changing the Balkans from a patchwork of pagan Slav kingdoms into an extension of Orthodox Christendom (Obolensky, 2000). Byzantine political and cultural life extended into the cold reaches of Russia and Scandinavia, where a string of archaeological finds of Byzantine coins and artefacts provide correlating evidence to support the primary sources' claims of Greek-Viking interaction (Benedikz, 2007). In Byzantium's elite classes, a taste for classical Greek culture was revived. This started out as a scholarly interest in finding and preserving books from the empire's Greek and Roman past (as typified in the work of Photius, whose list of important books to read is the sole evidence for dozens of otherwise lost works – see Diller, 1962), but soon turned into a mini-renaissance of Hellenism. The Byzantines were aware of their Greekness: although they called themselves Romans, they knew they were the heirs of a classical Greek world that favoured philosophy, literature and history as the leisure pursuits of leisured gentlemen (Mango, 1980). In this third section of the chapter, I will use the writing of Constantine Porphyrogenitus, Psellus and Anna Comnena to chart the re-emergence of Byzantine power and the ritualized, gendered leisure of the empire in the 900s and 1000s.

Constantine VII Porphyrogenitus was emperor in from 919 to 959 but spent a large part of his reign a virtual prisoner in the imperial palace (Toynbee, 1973). This imprisonment by his regents and co-emperor is evidently partly to explain why he turned to history and writing: with no decision to make, and an endless leisure time to fill, something had to take over his communicative desires. Instead of resorting to sex and drink, as so many other lost princes and caliphs, Constantine found solace in compiling and synthesizing a number of books about the ritual and political life of the imperial court. What survives of this work demonstrates that the imperial court became transfixed by ritual and routine. Much like the Groans of Mervyn Peake's Gormenghast (Peake, 1946, 1950), and their castle attendants, life in Byzantium revolved around a carefully stage-managed processional, with emperors making public visits through the city and the hippodrome (Gregory, 2005), receiving ambassadors in state rooms and having servants even in the most private of leisure spaces (Toynbee, 1973; Mango, 2000). The entire public life of the city turned on the imperial palace – as such, everyday leisure lives were constrained by the demands of the rituals: taxes were raised to pay for mechanical objects sent as presents to the Franks in the West (Gregory, 2005), service industries thrived but with long working hours, and taverns, baths and brothels were subject to arbitrary control or closure to ensure that no imperial personage – or their guests – were tainted by the sin those places endowed on the pious Christian (Mango, 2000).

Michael Psellus (c. 1017–96) was from a middle-ranking family of the Byzantine Empire, which offered him private wealth and the luxury of a full education in the Hellenist culture revived a century earlier.

He was a close advisor of a number of eleventh-century emperors and was closely aligned to the aristocratic elite of the city (Hussey, 1935). He wrote a history of his times, the *Chronographia*, which follows the classical tradition of listing the virtues and vices of the great figures in imperial life: eunuchs, emperors, men, women, advisors, generals and usurpers. Psellus is recognized by historians as an important, and mainly accurate, primary source (Gregory, 2005), although, of course, he shows some factional and personal biases (Hussey, 1935), not least in the absence of detail about the life of the empire's poor masses. The history is invaluable as a record of everyday life and leisure habits, as well as political life, as Psellus carefully observes both virtues and vices. For instance, in his long description of the wastrel Emperor Constantine VIII, brother of the more competent Emperor Basil II (2: 7–9, English translation by Sewter, edited by Halsall, on-line at http://www.fordham.edu/halsall/basis/psellus-chrono02.html) he writes:

> He was especially expert in the art of preparing rich savoury sauces, giving the dishes character by combinations of colour and perfume, and summoning all Nature to his aid – anything to excite the palate. Being dominated by his gluttony and sexual passions, he became afflicted with arthritis, and worse still, his feet gave him such trouble that he was unable to walk. That is why, after his accession, no one saw him attempt to walk with any confidence: he used to ride on horseback, in safety. For the theatre and horse-racing he had an absolute obsession. To Constantine these things were a matter of real concern, as he changed horses, harnessed fresh mounts, and anxiously kept his eye on the starting-points in the arena. The *gymnepodia*, long ago neglected, was also revived in his reign. He reintroduced it into the theatre, not content with the emperor's normal role of spectator, but himself appearing as a combatant, with opponents. It was his wish, moreover, that his rivals should not be vanquished simply because he was the emperor, but he liked them to fight back with skill – his own credit for the victory would then be the greater. He used to chatter away, too, about his contests, and he mixed well with the ordinary people. The theatre also attracted him, and no less the chase. Once engaged in the latter he was impervious to heat, ignored the cold and never gave way to thirst. Most of all he was skilled in fighting with wild beasts, and it was because of this that he learned to shoot with the bow, hurl the javelin, draw his sword with dexterity, and aim his arrow straight at the mark. He neglected the affairs of his Empire as much as he devoted himself to his checkers and

> dice, for so ardent was he in the pursuit of gaming and so enraptured by it, that even when ambassadors were waiting to attend on him, he would neglect them if he was in the middle of a game. He would despise matters of the utmost importance, spend whole nights and days at it, and fast completely, voracious eater though he was, when he wanted to play at the dice.

This extract from Psellus shows that in the tenth century CE (the years in which Constantine VIII ruled) and the eleventh (when Psellus wrote this), there was still, in Byzantium, an instrumental role for games, athletic contests and the theatre. Spectacles still helped maintain order. Furthermore, the private lives of the rich were – despite the close profession of Christianity – all too easily turned to dark leisure practices such as gluttony, sex and gambling. All these things must have existed in Psellus' day and must have been familiar enough to his readers: either in their own lives or in those of the masses in the city around them. Evidence for this is found elsewhere in the book, when Psellus draws metaphors from leisure life to explain the personalities of individuals or power relationships in the court. For example, he writes this about the Chamberlain Basil, who effectively runs the state while his young Emperor Basil II grows into responsibility and adulthood (1:3, English translation by Sewter, edited by Halsall, on-line at http://www.fordham.edu/halsall/basis/psellus-chrono01.html):

> The *parakoimomenus* [the Chamberlain Basil], in fact, was like an athlete competing at the games while Basil the emperor watched him as a spectator, not a spectator present merely to cheer on the victor, but rather one who trained himself in the running and took part in the contests himself, following in the other's footsteps and imitating his style. So the *parakoimomenus* had the whole world at his feet.

Later on in the history, when summarizing events he had seen with his own eyes – the increasing despotic rule of Theodora, who married a series of men in an attempt to keep power; and the successful rebellion of Isaac Comnenus (reigned 1057–1059) – Psellus imagines the empire as a chariot being driven by different emperors in a race (7: 56–58, English translation by Sewter, edited by Halsall, on-line at http://www.fordham.edu/halsall/basis/psellus-chrono07.html):

> Theodora's drama played out to its finish, the reins were put into the hands of the old man Michael. Unable to bear the movement of the

> imperial chariot, with his horses running away with him from the start, he made the show more confused than ever, and being scared out of his wits at the uproar, he retired from the race and took his place by the non-runners. Of course he ought to have held on; he should have kept a pretty tight hold on the bit. In practice, however, he was like a man who is dismissed the service – in his case, the throne – and returns to his former manner of life…The opportunity for healing recurred and Isaac Comnenus, wearing his crown, climbed into the Roman chariot…Isaac was a devotee of the philosophic life: he abhorred anything that was physically diseased or corrupt. But his hopes were disappointed, for he found nothing but disease and festering sores, the imperial horses running at full speed from the starting-post, quite impossible to master, heedless of the reins.

Anna Comnena (1083–1153) was the daughter of an Emperor, Alexius I, who reigned from 1081 until 1118. Her *Alexiad* records his life and deeds as well as her own thoughts and observations of matters in her day. The Alexiad is important as it records the arrival of the first crusaders to Constantinople, travelling on the pilgrim trail to the Holy Land to try to defeat the Muslims. To Anna Comnena, the Franks, the Western crusader-pilgrims, are barbarians, illiterates with a lust for base activities in their leisure time. She contrasts their uncouth ways with the careful ceremony and ritual of her father's court and the Christian piety of Constantinople's citizens. She describes the Franks (or Kelts, as she calls them) as 'brazen-faced, violent men, money-grubbers and where their personal desires are concerned quite immoderate' (Comnena, 2003, p. 450). This allusion is as close as Anna Comnena gets to the dark leisure habits of the foreigners that contrast with the (undoubtedly hagiographic) heroic Christian leisure virtues of her father.

Enter the Turks

As well as providing a first-hand account of the arrival of the Crusading Franks, the *Alexiad* also tells of rise to prominence of the Muslim Turks, who very quickly replaced the Byzantines in Asia Minor. In this final section, Byzantine leisure lives will be compared and contrasted with the leisure lives of the Ottomans, who succeeded to power in Constantinople. Secured in the highlands of modern-day Turkey, the Turks contested power with their Christian rivals over a number of centuries following their defeat of Emperor Romanos IV Diogenes in 1071 at Manzikert. The idea of Byzantine decline following that battle has been contested by historians and archaeologists who point to a number of resurgences of Byzantine power and culture, especially in the years immediately following the re-conquest of Constantinople in 1261 from the Latins (Gregory, 2005). However, it is a matter of fact that the Turks, under the leadership of the Ottomans, conquered all of Asia Minor by the end of the fifteenth century, along with most of the former Byzantine provinces of the Balkans (Harris, 2010). Constantinople was left unconquered for another fifty years, along with a handful of other cities and regions – but even Constantinople fell, at last, to the army of Sultan Mehmed II in 1453.

After the fall of Constantinople, Western Europeans developed and maintained embassies to the Ottomans, as they had done with the Byzantines. In reports sent back West by these ambassadors, and by books published by visiting traders, Western prejudices about the tyrannical and irrational Other were often perpetuated (Yapp, 1992). But Yapp (1992, pp. 148–49) also describes how more positive opinions of the Ottomans were transmitted:

> The good image was common among some of those who had a close acquaintance with the Ottoman empire, and was...offered most memorably by the imperial ambassador, Ogier Ghislain de Busbecq, in his *Turcicaele gationes epistolae quatuor* (1589), a book which contrived to depict the *devshirme* as the equivalent of success in the civil-service examination. Busbecq's image was of a people very different from that of Europe, but nevertheless possessing its own admirable qualities: military and administrative skills, a tolerant government, a system of justice which was simpler, quicker and less corrupt than that of Europe, a stable social order and pleasing personal characteristics: endurance, frugality, sobriety, cleanliness, politeness and hospitality.

What is clear from Busbecq's letters is how so much cultural life survived from the Byzantine world into the Ottoman one (Babinger, 1992). What is notable about the cultural shift from Byzantine Christianity to Turkish Islam is clear continuity of leisure practices from the conquered to the conquerors. Even before the fall of Constantinople, the Ottomans had adopted much of the ceremonial and political culture of the Byzantine elite (Inalcik, 2000). With this high culture came a taste for bathing and music, to complement the Turkish appreciation of poetry and hunting (Faroqhi, 2005; Goodwin, 2008). Sports and games were still common-place, played during public festivities but controlled by the guilds that enacted imperial and theological constraints on individual leisure lives (Goodwin, 2008). Ottoman rulers learned to support these events – as good ways of ensuring the armies were battle-ready, and as good ways for releasing everyday tensions among the lower and middle classes. Alongside the instrumental nature of public ceremonial, in the years following the conquest of Constantinople, the Ottomans also adopted the private excesses of the Byzantine imperial palace: the keeping of mistresses (latterly in harems); the consumption of huge quantities of food; and the drinking of wine in such an excess that certain Ottoman rulers such as Selim II (reigned 1566–74) became notorious in Islamic history for their insobriety (he is nicknamed the Drunkard: see Goodwin, 2008).

Away from the elite, leisure spaces in Ottoman Constantinople were not very dissimilar from Byzantine leisure spaces. For the minority Christians in the Ottoman Empire, taverns were still a source of alcohol. For the Muslim majority, coffee-drinking quickly became established as the equivalent of alcohol-drinking: coffee houses opened up where men could sit and drink with other men in the evening, to talk, and to play games such as backgammon (Faroqhi, 2005). Public baths remained open and used, but again the clientele was male, that gender which was allowed access to the public sphere. Music continued to be played in Greek scales, and although the Ottoman Empire saw a number of movements arguing for Muslim orthopraxy over the un-Islamic nature of music, music (and dance) continued to figure in people's everyday leisure lives (Inalcik, 2000). Away from the hegemonic gaze of the imperial centre – in the dark streets of Constantinople, on the plains of Cappadocia, in the mountains of Albania – alternative leisure lives continued to be played out by the irreligious, the immoral, the criminal or anyone else looking to find quick satisfaction and release from the grim pressures of life. Prostitutes offered their services, men gambled on the throw of dice, men fought each other, the unscrupulous bought and sold goods on the black market, and although alcohol was not allowed, Muslim men bought and drank wine and spirits (Barkey, 2008).

The Ottomans did differ in some ways to the Byzantines in their leisure practices: the pilgrim pursuit of the *hajj*, for example, was supported by the Ottoman control of the entire region from Asia Minor through Palestine and Syria to the holy cities in Arabia (Faroqhi, 2005). Furthermore, the Ottomans were content to allow cultural differences across separate provinces, which allowed Egyptians to retain feasts to celebrate the Nile floods and Shi'ites to venerate their martyrs, and gave Christians and Jews much autonomy over the governance and religious practices of their communities (Barkey, 2008). Sufis found space to find God through dancing (Faroqhi, 2005). This multicultural tolerance allowed local leisure practices to flourish away from the immediate control of the imperial palace, which made the Ottoman Empire quite different from the Byzantine.

However, the similarities outweighed the differences. There was, then, a Byzantine leisure inheritance into the Ottoman Empire that worked on a number of levels. Both states routinely used leisure in similar instrumental ways, to ensure the continuity of political stability, communal identity and inequalities of power. At the same time, the urban nature of the centre of power ensured that alternative, communicative expressions of leisure continued to be found in informal and commercial social exchanges.

6
Leisure in the Middle Ages

The chronological order is continued in this chapter, and leisure in the European Middle Ages is explored, drawing on the relationship between Christendom and Islam and arguing for the influence of the latter on the former's courtly leisure culture (Beaver, 2008). The first section of this chapter will give an overview of the rise of feudalism from the Merovingian and Carolingian Empires through to the 1200s, with discussion on the connection between feudal relationships and formal leisure. I will argue that the rise of the manorial village, with its complex and ritualized culture and calendar, constrained the leisure choices of both the peasant and the lord. I will discuss the prevalence of folk games such as football and will compare the leisure lives of the peasantry with the world of feasts and jousts of the feudal princes, and will argue, following Habermas, that the latter world was limited in its expressive, communicative freedoms. As a case study in understanding the past, I will discuss the Vikings and evidence for their leisure lives. Medieval Islam will be the focus of the second section of this chapter, where it will be argued that freedoms associated with pilgrim trips, mercantilism and a tradition of philosophical education were constrained by the arbitrary power of political and religious rulers. The relationship between Islam and Christendom will be explored in this section, and in particular I will show that the rise of hunting, falconry and romantic court music as elite leisure forms was due to the Arabization of Christian princes in Sicily, Palestine and southern Italy. In the third section of this chapter, I will follow this discussion of the Mediterranean world with an account of the rise and fall of the troubadours in the south of France. I will argue that the troubadours were an expression of political and religious freedoms across classes, and that the decline of troubadourism was inevitable once the Cathar world in which they lived was extinguished by the Albigensian Crusade. In the final section of this chapter, I will begin with the example of the sea-shell badge mentioned in Langland's *Piers Plowman* to discuss pilgrimage as a nascent tourism industry, using the examples of Rome, Jerusalem, Santiago, Canterbury and Mecca.

Feasting in the construction of community

When the empire collapsed in the West, and transformed itself in the East into the Christian city-state of Byzantium, it was inevitable that an echo or memory of fifty-century Athens survived into the Christian Europe of the early Middle Ages. But Europe in the Middle Ages was also shaped by the norms and values of the barbarians, who replaced the Roman Empire (Heather, 1998). Discourse was replaced by highly stylized shows of strength and power. In her book *Creating Community with Food and Drink in Merovingian Gaul*, Effros (2002) explores the centrality of the feast in the leisure lives of the people of the old parts of the Western Roman Empire in the fifth and sixth centuries CE. In the aftermath of her decline of the Western Roman Empire, two seemingly divergent cultures appeared in what is now France: the Gallo-Romans, the heirs of the Classical tradition, who clung to an urban life and the Latin of school and church; and the barbarians, mainly the politically dominant Franks who retained traditions of military strength, masculine prowess, paganism and the ideal of the Germanic war-band (Wood, 1993). The identity of these two groups is, of course, highly disputed. The traditional theory that the Franks were Germanic invaders – who simply defeated the Romans and migrated over the Rhine en masse (typified by the writing of Stenton, 1970) – has been questioned by more recent historians. Some, such as Halsall (2007), argue that the Franks emerged out of units of the late Roman military, units of federates, auxiliaries and Romans adopting Frankish traditions. Wickham (2010) argues that the divide between Frank and Gallo-Roman was essentially a fluid one, and marked more by cultural preferences than supposed ancestors. Heather (2006) acknowledges these cultural arguments but still returns to the fact that primary sources written at the time, such as Sidonius and Gregory of Tours, argue that there were clear differences – linguistic, political, and religious – between civilized Roman and barbarian Frank. The cultures do not seem permeable from the perspective of Gallo-Roman sources, but, of course, those sources are from the class of literate, city-dwelling elites who were marginalized by the warrior-masculinity of the Frankish aristocracy. Effros' (2002) work on feasting demonstrates how the two communities of Merovingian Gaul were brought together in a common community, formed through the giving of feasts by the Frankish lords, and the receiving of such largesse by lesser lords, tenants and urban-dwelling Gallo-Romans. This culture of the feast at the centre of a complex relationship of obligations was something shared in common between the Franks and the Gallo-Romans: wealthy lords from the latter replicated the custom, so that it became something shared across the entire community, a leisure pastime loaded with communicative meaning and instrumental purpose.

The Merovingian Empire, and its client relationships, was the originator of what came to be known as feudalism (Wood, 1993). The problem of land – and tenure of rights and obligations – fed into an increasingly hegemonic control at the centre of the Frankish Empire, which, in turn, led to a cycle of revolts and rebellions on the periphery (Halsall, 2007). Under Charlemagne, the Frankish Empire became, briefly, a new Roman Empire (McKitterick, 2008; Wickham, 2010), but one based on absolute power and quasi-feudal relationships. Charlemagne could not physically control the huge domains he conquered. Instead, he had to appoint lords who ruled for him, autonomously, but who held their lands in his name. These lords were expected to provide military support for Charlemagne, and money in the form of various tithes. In turn, these lords were given the freedom to sublease portions of their fiefdoms to lesser lords, who, in turn, sublet their portions to knights – free men with military obligations – or peasant tenants. The tenants at the bottom of the pyramid were effectively slaves to their lord, even if they were legally free (Wickham, 2010). This system – the feudal system – differed in many places in Western Europe, but in essence the system that emerged under the Carolingians was the one adopted by Western Europeans wherever the Roman law of citizenship had been replaced by laws based on barbarian or pre-Roman moral codes of masculine honour and the complete submission of inferiors (McKitterick, 2008). This system formed the basis of the law and popular customs in Norman England, in Normandy and France, in many parts of Germany, and Lombardy and Austria, until the thirteenth-century, when a surge of urbanization allowed town-dwellers the chance to be free of feudal ties (Moore, 2000).

In describing the way in which the public manifested itself in post-Roman and medieval times, Habermas (1989[1962]) identifies leisure as a site for such representation and reproduction. Feudalism was about the show of power, and the tight leash of control over everyday lives. The rise of the manorial village, with its complex and ritualized culture and calendar, constrained the leisure choices of both the peasant and the lord (Dyer, 1994). For the peasants, life was dictated by the cycle of the seasons, and the holy days of the Church. Such holy days would often allow some freedom from the toil of agricultural work, and the feasts associated with them were often drunken and debauched affairs (Masschaele, 2002), but the holy days brought the peasant and lord together to take part in formal processions and services presided over by the local church. There is some evidence that holy days sometimes allowed the poor to transgress norms of feudal society: having sex in the woods, or playing at king for a day (Petitt, 1984). However, that evidence is not evidence of normal practices, and most medieval historians and archaeologists would argue that holy days were treated seriously by all sections of feudal society, and the leisure time of these days was spent reinforcing the norms and values of feudalism and Christianity (Masschaele, 2002). One of the leisure pastimes that is recorded as occurring during holy days are the folk games held among different manors, parishes or villages. The most famous is, of course, football, but foot-races, horse-races and other competitions were held across Europe (Guttmann, 1981).

Early football was, by all accounts, a riotous affair (Holt, 1989; Strutt, 1969[1801]; Williams, 1994). Dunning and Sheard (1979) chart the problems that this football caused to the daily life of the villages and populations of pre-nineteenth-century Britain, and see in the unstructured games a chaotic violence that was tamed by the civilizing process in the nineteenth century. But these games were events, part of the celebrations of holy days, and what to Dunning and Sheard and others may appear to be chaos, to the participants of the time it was probably framed within a system of tacit or verbally agreed rules (such as the Cornish game of hurling[1]). Although the game was proscribed by a number of edicts, this did nothing to stop the popularity of the various types of game that were called 'football'. The change to a more efficient law-enforcing society based around the early industrial cities gave the anti-football establishment greater scope for controlling the game, and in 1835 the Highways Act banned the playing of football on public highways, removing the traditional arena of the game. Yet, folk football continued in a few places into the latter half of the century (Williams, 1994).

Traditional football is still played on holy days in parts of the country in England, though there is a question mark over whether these are invented traditions, modern creations and behaviours that are linked with, and justified by, the past through a false sense of continuity (Hobsbawm and Ranger, 1983). What mattered with folk football and other games was that men from the poor classes could find a way to demonstrate their masculinity through some athletic prowess. Beating the neighbours across the parish boundary was a way of bringing honour to one's local community, proving to the lord one's loyalty to the fief, proving perhaps to a girl one's manhood and virility (Lees, 1994). Such competitions were also a way of encouraging drinking, gambling and interaction among different communities and were often associated on the holy days with markets and fairs (Masschaele, 2002).

Alongside the feudal festivals associated with the manor, more formal feasts became associated with the ruling classes of Western Europe (Dyer, 1989; Woolgar et al., 2009). The feast itself became more elitist in the eleventh century onwards, reflecting the decline of the communitarian tradition of earlier years, and the ossification of high feudal culture. No longer were free men of all ranks invited to eat at their lord's table – instead, the feast turned into the medieval banquet, where huge amounts of meat and drink were consumed to demonstrate wealth and ostentation (Dyer, 1989). Around these banquets developed a plethora of service roles: cooks, serving men, musicians and dancers (Hammond, 2005). The feasts themselves became punctuations to more formal leisure interactions: hunting expeditions on land set aside for such chases (Almond, 2003) and the knightly pastimes of combat and jousting (Crouch, 2006).

Habermas explains how the joust sublimated an ideal of public nobility that was different to the ideal of nobility in Ancient Greece, but which was still predicated on the sleight of hand of hegemonic power. For Habermas (1989[1962], p. 8) the joust was linked to the staging and representation of power, expressed through the rituals of the performances and the symbolism and rhetoric of culture and status. The joust was the visual representation of a hierarchy of power culminating in the good Christian knight, the noble lord holding land; jousts were not only expressions of good courtly behaviour, but also, and more importantly, allowed the participants and spectators to share in a celebration of a holy day and the hegemonic masculinity of the chivalric ideal:

> Steel that he was, his courage never failed him, his conquering hand seized many a glorious prize when he came to battle... Thus I salute the hero. – Sweet balm to woman's eyes, yet woman's heart's disease! (Wolfram von Eschenbach, 1980, p. 16)

In the thirteenth-century poem *Parzival* cited above, the aspiring knight is given an image of what it means to be a man, to be a knight. The hegemonic masculinity portrayed is one of nobility, courage and prowess in battle. However, the site of developing and reinforcing this masculinity is not found in war: instead, it is the tourney ground where the allegorical character of the perfect thirteenth-century noble man is found. In comparing the leisure lives of the peasantry with the world of feasts and jousts of the feudal princes, we can see, following Habermas, that the latter world was limited in its expressive, communicative freedoms. In the freedom of a house, peasant men and women could find some private, domestic leisure activity to satisfy their communicative desires, whether each other's company, or in drinking and telling stories that mocked their masters in the manor house (Dyer, 1994). For those members of the male nobility forced into the cycle of attending their superiors and proving their worth as knights, their was the a constant pressure to meet those expectations with a meek smile (Almond, 2003); for noble women, one's leisure life was constrained by one's value as a bride, or the expectation one maintained a blood-line (Berg and Power, 1997). There were exceptions to all these rules, but the essential differences remained in Western Europe wherever feudalism was firmly entrenched. However, on the fringes of the feudal world, some additional communicative freedoms remained.

For instance, Icelandic Sagas survive from the Middle Ages that tell us of the earlier world of the Vikings – they are oral stories, part history, part fable, written down by Christian monks. *Njal's Saga* is typical of the sagas (Cook, 2006). These sagas tell us of the leisure lives of the Vikings – which, in turn, relates to how Vikings are represented in modern leisure forms, such as sport, music and tourism (von Helden, 2010). *Njal's Saga,* for instance, tells us of the value attached to brave deeds, the importance of gifts and of song in formal leisure rituals, the heathen traditions of the Vikings and the survival of these traditions after the coming of Christianity, the commonwealth and early history of Iceland, the 'pagan' nature of morals and values, the continuity of beliefs in magic and elves into the Middle Ages, and the importance of extended families, private leisured spaces and public oaths of loyalty. But it also tells us of the violence of men: Njal is an honest man who is killed because of his refusal to disown his sons. The Sagas present a 'modern' view of women – independent-minded, in charge of households, with many freedoms (divorce, friendship, free time and land ownership). Archaeology backs up the view of independent Viking women – for example, in the rich burial goods found in women's graves (Jochens, 1998). But the Sagas also paint a stereotype of women: that they are not to be trusted, that they scheme behind the backs of their husbands.

The Sagas tell of visits to the north of England and York (Jorvik). Despite the short amount of time that York spent under Viking influence (about a hundred years) compared to Saxons (about three-hundred), Britons (two hundred) and Romans (four hundred), its history as Jorvik provides its most visible historical identity. York's Vikings were politically and culturally related to the Norse colonies in the Orkneys and in Ireland, other places mentioned in the Sagas, where curious Norse visitors have carved their names at the Neolithic site at Newgrange in Ireland (Jones, 1984), in the same way that modern-day tourists write their names at the top of the Empire State Building.

The Jorvik excavations are a major tourist destination; major archaeological site for everyday life in tenth-century Viking York. The Jorvik Centre is more than a so-called tourist destination, it has become a representation of a certain parochial Yorkshire independence (Ehland, 2007). Viking Yorkshire is seen as different to Saxon southerners. Vikingness in leisure is something 'desired' by modern-day chauvinists across the world, from the pagan black metal music scene in the Ukraine (von Helden, 2010) to a plethora of sports teams in North America, Australia and England with the Viking as their nickname and mascot. Vikings, like the Vandals, are now used as shorthand for hooligans. Vikings are seen as fighters – negatively, as brutes, positively, as noble warriors. The 'horned helmet' has passed into representations of Vikings, even though they never wore them (nineteenth-century Romantics depicted Vikings in 'Roman' helmets). Historians (e.g., Jones, 1984) now see the Vikings as being a vital, creative and dynamic culture, even if others, such as Nazis, neo-pagans and beer companies, have appropriated the Vikings for their own purposes.

Medieval Islam and its influence on Western Europe

Medieval Islam, apart from the form that flourished under the Ottomans, vacillated between freedom of thought and religious conservatism. The bloody political and social history of Spain provides the clearest exemplar of this tradition of freedom and constraint (Fletcher, 2001). Under certain rulers and ruling families, Muslims, Jews and Christians mingled in the public sphere and established a unique multicultural and multifaith culture that was brutally repressed the by the winning Christians after the fall of Granada in 1492. This culture operated at elite and popular levels and was a site for and exchange of ideas about leisure practices and activities.

At elite levels, there was a convergence between Islam and Christendom over what was considered 'proper' for a prince in his everyday public and private life, including his leisure choices (Runciman, 1992; Lowney, 2006). A true noble was expected to be a gracious and generous host, a lover of hunting and falconry, and a protector of travellers, scholars and various minstrels. He was supposed to encourage and support mercantile adventures, though he himself was too noble to be directly involved in such buying and selling. In private, a true noble man was expected to be a passionate and chivalrous lover.

At the level of popular culture, Muslim Spanish traditions were also adopted into Christendom: a love of story-telling, poetry and bull-fighting, which survived from Roman times through the Islamic period (Douglass, 1999; Fletcher, 2001). However, the freedoms associated with travel, mercantilism and a tradition of philosophical education were constrained by the arbitrary power of political and religious rulers (Kennedy, 1996; Hourani, 2005). In Spain, the periods of multicultural tolerance were interspersed between periods of conservative retrenchments, which limited the freedoms of Jews and Christians and enacted strong punishments for un-Islamic activities, such as drinking alcohol (Lowney, 2006).

The relationship between Islam and Christendom in the Middle Ages is a crucial one for this history of leisure. What we see is an elite convergence, where the communicative expression of nobility is codified in Medieval Islam, which is adopted by Christians in the liminal lands of southern Europe and Outremer (Syria and Palestine – see Fletcher, 2004). This communicative expression of the nobility is expressed through the taste for particular leisure forms. Following Bourdieu (1986), we could say this was an exchange of a particular form of cultural capital as leisure, which was used to create the *habitus* of the medieval Christian prince (Runciman, 1992). In particular, the rise of hunting, falconry and romantic court music as elite leisure forms was due to the Arabization of Christian princes in Sicily, Palestine and southern Italy (Karabell, 2007).

One site of this exchange and convergence was the Middle East, where Christian crusaders carved out the fiefdoms of Outremer from the *umma* of the Muslim world. Before the arrival of these Christian warlords, the Muslim Arabs of Syria and Palestine lived cultured and cultivated lifestyles, prosperous with the profits of trade and industry, and relatively free from autocratic rule (Maalouf, 1984; Fletcher, 2004; Hourani, 2005). They lived alongside and mingled with their Christian and Jewish neighbours, and preserved some of the cultured leisure forms of their Ummayad and Roman ancestors (Ansary, 2010). Maalouf (1984) reports on the confusion and bewilderment expressed in Arab sources at the arrival of the uncouth, rude and boorish Franks. Rather like the Byzantines at the time, the Muslim Arabs saw the Franks as irrational and overly-aggressive, and driven by their passions into debauched leisure. But once the Christians had achieved their aim of capturing Jerusalem, the fiefdoms of Outremer became places where Christian brutishness was polished by education and interaction with the locals: the Franks who remained behind – or who were born in the East – learned and adopted the Muslim norms and values on hospitality, the wisdom of learning, a love of hunting and falconry, and moderation in all things (Ansary, 2010). The Knights Templar and Hospitaller, for example, were clear cases of Christian elites mirroring local elites in the protection of pilgrims and other travellers, and providing generous hospitality to strangers (Fletcher, 2004). Although these militaristic holy orders later emerged as important landowners and financiers back in Christian Europe (which led to the demise of the Templars – see Read, 2003), they still maintained rules that provided guests food and lodgings, and still practised the leisurely, but noble, arts of hunting and falconry.

Falconry is an interesting example of elite convergence. Maintaining hunting birds was an expensive task: the trade in such birds was expensive, as was their training, and skilled keepers were hard to find (Oggins, 2004). Feudal lords, however, were fearful of hunting birds being kept by the lower orders, so customs and laws emerged that limited certain birds to certain ranks in the social order (Glasier, 1998). The elite rules governing birds-of-prey came ultimately from Islamic Sicily, Palestine and Spain. It comes as no surprise, then, to find the transfer of knowledge and culture from Islam to Christendom on birds-of-prey occurring in those areas (Runciman, 1992; Ansary, 2010). In the 1240s, King Frederick II of Sicily even wrote a book on the subject, the *De arte venandi cum avibus*, or *The Art of Hunting with Falcons*, in which he wrote:

> Although it is true that birds of prey display and inborn antipathy to the presence and company of mankind, yet by means of this noble art one may learn how to overcome this natural aversion...Here it may again be claimed that, since many nobles and but few of the lower rank learn and carefully pursue this art, one may properly conclude that it is intrinsically and aristocratic sport; and one may once more add that it is nobler, more worthy than, and superior to other types of venery. (Frederick II, translated in Fyfe, 1943, p. 6)

Frederick was sympathetic to Islam and Islamic culture: he had many Muslim advisors in his court; he hired Muslim architects and scholars and expressed an understanding of Islam that made many of his contemporary rivals question his Christianity (Runciman, 1992). Frederick followed guidelines on how to be a proper prince, which he inherited from the earlier Muslim rulers of Sicily: he encouraged the study of medicine and philosophy; he encouraged the translation of Islamic books; and in private, he surrounded himself with the luxuries of wealth (Ansary, 2010). All of these things made Frederick's Sicily a multicultural island, where popular and elite leisure forms were subject to free exchange of ideas, entrepreneurial speculation and communicative choice.

Troubadours and the Holy Grail

In this third section of the chapter, I will follow this discussion of the Mediterranean world with an account of the rise and fall of the troubadours in the south of France. I will argue that the troubadours were an expression of political and religious freedoms across classes, and that the decline of troubadoureanism was inevitable once the Cathar world in which they lived was extinguished by the Albigensian Crusade. Before I discuss troubadoureanism, it is important to put their poems in the context of the wider poetry and literature of the Romance movement. This Romance movement was a twelfth-century borrowing from Islamic Spain (Lowney, 2006), although the context of the Christianity shaped the image of the chivalric knights and their pious (but loving) ladies (Solopova and Lee, 2007). Famously, the Arthurian tales of courtly love, of jousts and honour, were reimagined as theological allegories of Edenic sin and resurrection (Echard, 2011). In this remaking and retelling, the Romance poems spread and popularize the idea of the Holy Grail, which, in turn, through the medium of the troubadours, became a familiar object in people's leisure lives: in the quiet of noble courts, or the more noisy yards of wayside inns, people listened to the stories and took on their instrumental lessons about the right way to be Christian and the chaste way to love (Carruthers, 2010).

The Holy Grail, as we know it, as it was first envisaged, is a Christian object (Echard, 2011). Remember, we are not dealing with mundane fabulousness, only in poetic metaphors. The Holy Grail exists but not in the physical wood-carved-cup way, which some people would have us believe in. The story of the Holy Grail and Glastonbury and Joseph of Arimathea comes after the event of the writing of the early anonymous thirteenth-century *Quest of the Holy Grail* (Matarasso, 1975) or *Parzival* – in these two texts, its mundane origins as the cup of Christ is not important or stated. Popular mythology has distorted and hidden the original meaning of the Holy Grail in the texts, and now we have people who believe that the Holy Grail is hidden in the attic of a West County cottage or on the mantelpiece of the Vatican. Copyists are never too concerned with capturing meaning, only images, and Malory knew a good story when he saw one, when three hundred years later he revived the Arthurian cycle as a repository of Englishness and English Christianity. The connecting of the cycle with Geoffrey of Monmouth's 1136 'history' of Britain (Geoffrey of Monmouth, 2004) equated Tudor England and its contemporary glory with the glories of Rome and – through the objectification of the Grail – with Christ Himself, was normalized and naturalized. And Malory's (2008[1470]) version of the Arthur story has become the origin point of the myth we are familiar with today, distilled through Anglo-Saxon romanticism typified by Tennyson, or Celtic pagan romanticism (even though the Celtic claim to fame is their brand of Christianity and its importance for much of the first millennium).

We must first turn to that errant knight Wolfram von Eschenbach, whose confused second-hand tale of the pure knight Parzival and his run-in with the Grail Family is the subject of much hermetic debate. It is true to say that von Eschenbach's work lives outside the mainstream of Arthurian romance, and there is little common ground between the mystical German story and the contemporary drama of the French troubadours typified by Chretien de Troyes. However, this makes the claim made by von Eschenbach – swallowed without doubt by some conspiracy-minded theorists (Eschard, 2011) – that his work is a rework of a French source rather insubstantial. Scholars agree that von Eschenbach, while being influenced by the French tales, was drawing upon older Teutonic and recent Biblical folklore to create a wondrous land through which Parzival is transformed from impure knight to Christ.

The early history of *Parzival* reflects the cosmopolitan nature of learned Europe and its links with Islam, and this is mirrored by the clever use of the Templars as a motif of mystery and purity. It comes as no surprise (to the thirteenth-century reader) to find the Keepers of the Grail described as templars, as the role of the real world in defining the world of the fantastic was common in what was still a mainly oral and social literature. It is a historiographical sin to assume that since they are called templars, they must be The Templars, and hence von Eschenbach is 'in' on secret knowledge that the Templars are in possession of the Grail or quasi-Grail artefact. If he was privy to such information, he would surely have kept it secret, not put it as the central theme in a story. Von Eschenbach, however, was interested in alchemy, and the alchemical route to God is the subject of his reworking of traditional stories and the early Arthurian cycle.

Parzival is seeking not the Grail but the Graal, an altogether different thing. It is described in the story as a stone, a life-giving stone that has fallen from heaven. That does not mean the stone is an asteroid or a space ship, but is to be read as a symbol for a thing of purity that exists in the alchemical quintessence, a thing made from the element of the heavens – a thing known to the medieval world from Islam as the *lapis exillis*, the Philosopher's Stone. Those who read *Parzival* in the period would have recognized the Graal for what von Eschenbach intended it to be – for to an alchemist, the route to the Philosopher's Stone was a route to God, a spiritual quest, where the gold to be found was gold for the soul, or for the world of the four elements (Grant, 1997).

So *Parzival* is aimed at an educated audience familiar with the Arthurian romances and alchemy, and makes the connection between the two for the edification of the reader or listener. It is a lesson in piety couched in the language of magic, which was not a supernatural thing but a thing of great purity – a path to spiritual one-ness with the mysteries of religion. The Grail is thus, in von Eschenbach, a symbol of the final state in which Parzival finds himself, once he had asked of its meaning, for its meaning is simply that of the quest itself.

Parzival provides a clue to the metaphor at the heart of the Grail, which we finally see in all its glory in the anonymous *Quest of the Holy Grail*, believed to have been written by a Cistercian monk at the beginning of the thirteenth century for his own leisure (Matarasso, 1975). This is a time when Cathar heretics have carved out an autonomous state in the south of France, and orthodoxy is threatened (Sumption, 1999). This is a time when Outremer, the Holy Land, is struggling and surrounded by the heathen Muslims. For a religious monk in northern France it would have seemed like the end of the world. What we see in the *Quest*, therefore, is the lifting of a well established motif, which is reworked into a beautiful and elegant message of Faith for the solace of believers in the dark times the monk believed Christendom to be in. Unlike *Parzival* and the Arthurian romances, the *Quest* is a holy text, expounding religious orthodoxy, interpreting the Bible and leading the listener through the Quest of Sir Galahad – Christ Himself – for what is called for the first time the *Sangreal*, the Holy Grail (Eschard, 2011). It is this that, in the final magical moments, transforms the pure knight Sir Galahad into an embodiment of the Godhood. He is, in the story, a metaphor for the Christian Soul, living in Christ. Hence the Holy Grail is not a simple magical cup: it has a far deeper meaning. It is the embodiment of Christian revelation, the medium through which God works between heaven and earth. This would be recognizable instantly by the religious audience of the time as being the Holy Spirit monad within the Trinity. Hence we have a complicated metaphor of Trinity between Sir Galahad, the Holy Grail and the Fisher King, with each taking on aspects of the Trinity in themselves, and the (Christian) importance of Christ central to each metaphor. So why is it called the Holy Grail? In the original Latin, the *Sangreal* provides the author with the chance to stress his metaphor further through a piece of wordplay common in medieval texts. For *Sangreal*, the Holy Grail, can be read through a different positioning of the space as *Sang Real*, the Holy Blood. This is the blood of Christ, the blood that is central to the Transubstantiation of the Eucharist, which is central to the Orthodoxy of Rome and anathema to the Cathar heretics. It is a joke, mocking the folklore of the cup, but at the same time it is a marvellous use of language to express political expediency and religious spirituality (Matarasso, 1975).

This digression through the reception of the Grail cycle is important, because the instrumentality of the various tales can be identified as a reaction against the increasing freedoms of leisure and thought in the small towns of the Languedoc in France (Sumption, 1999). This was where the troubadour movement emerged, as a combination of education, a loosening of feudal controls and political freedom amongst the free men and women of the towns, provided a space for poetry and music to be seen as central to everyday leisure lives (Paterson, 1995). The troubadours were a mixture of independent itinerants paid for their music, and noble amateurs dabbling in poetry: as in modern sports, these distinctions mattered to the individuals and their immediate social circles, but not to the consumers of the songs. For those entertained by the songs, troubadoureanism was new, exciting and dangerous (Shepard, White and Bruckner, 2000). It was the rap music of its day, seen as subversive by some in authority (especially the Church) but ultimately a musical form co-opted to the causes of the powerful (in the case of the troubadours, exemplified by the poetry supporting the Crusades – see Paterson, 1995). In their pure, communicative states, the troubadours took the themes of freedom and developed poems that explored the agency of true, romantic love, and bodily lust (taken from an anonymous poem, cited at http://en.wikipedia.org/wiki/Alba_(poetry)):

Quan lo rosinhols escria
ab sa part la nueg el dia,

yeu suy ab ma bell'amia
jos la flor,
tro la gaita de la tor
escria: "Drutz, al levar!
Quiieu vey l'alba el jorn clar
(While the nightingale sings,
both night and day,
I am with my beautiful
beneath the flowers,
until our sentry from the tower
cries: "Lovers, get up!
for I clearly see the sunrise and the day)

Troubadoureanism flourished in the Occitan language, spoken in the small towns of southern France. As mentioned, these small towns were also the heartlands of the Cathars, a religious movement banned as heretics by the Roman Catholic Church (Sumption, 1999). The Cathars believed that this world was evil, created by an evil demiurge – as such, things of the flesh were evil, and only spirit was pure. For some Cathars, this meant abstaining from most medieval leisure practices, such as eating meat, drinking and fornicating. For other Cathars, the evil of the physical world gave them permission to indulge in pleasures of the flesh as such pleasures could not harm their immortal and pure spirits. Cathars preached against authority and for freedom of expression – as such, they were a product of the same communicative popular culture as the troubadours (Paterson, 1995). Both groups travelled from town to town, protected by the free people and lords of each town, and given space in the towns to preach or sing. Some of the poetry of the troubadours even has Cathar influences (Shepard et al., 2009).

The troubadour movement spread from the south of France across Europe, where some of the songs and music were picked up by bands of wandering minstrels, which complemented the more populist and vulgar tunes they had for the entertainment of Northern Europeans (Haines, 2004). This transfer of troubadoureanism resulted in both a dilution of the romantic and spiritual themes of the songs, and a loss of the freedom of creativity to the control of the market. The troubadours were an expression of political and religious freedoms across all classes in the south of France, and the decline of troubadoureanism was inevitable once the Cathar world in which they lived was extinguished by the Albigensian Crusade.

Tourism and pilgrimage in Christendom and Islam

In William Langland's *Piers Plowman,* written in the fourteenth century, there is a shocking account of the medieval pilgrimage (Langland, 1978, pp. 2–3):

> Pilgrymes and palmeres plighten hem togidere
> For to seken Seint Jame and seintes at Rome;
> Wenten forth in hire wey with many wise tales,
> And hadden leve to lyen al hire lif after.
> I seigh somme that seiden thei hadde ysought seintes:
> To ech a tale that thei tolde hire tonge was tempred to lye
> Moore than to seye sooth, it semed bi hire speche.
> Heremytes on an heep with hoked staves,
> Wenten to Walsyngham--and hire wenches after:
> Grete lobies and longe that lothe were to swynke
> Clothed hem in copes to ben knowen from othere,
> And shopen hem heremytes hire ese to have.

Langland's poem is not merely a work of beauty – it is a unique window onto the everyday society and culture of late medieval England and Europe (Du Boulay, 1991). In it, we see a landscape famously described as a 'fair feeld ful of folk' (Langland, 1978, p. 1), a country where Christianity marks the movement of seasons, the round of holy days, and the day-to-life exchanges among the different classes (Webb, 2002). Langland's poem tells of the vision of a humble Christian, Piers Plowman. It is a reformist vision, which rails against corruption of the Church and the sinfulness of the elite (Baker, 1980): in terms of leisure practices, Langland views the sexuality, debauchery and gluttony of the higher classes with suspicion and horror (Zeeman, 2006). However, he is not a revolutionary: he wants the social order to be the way it used to be, where everybody knew their place. In essence, Langland wishes for a utopian past, a not-quite-perfect but perfectly Christian feudal society that existed before the Black Death damaged the feudal relations that typified the earlier Middle Ages (Huppert, 1998). It is a form of nostalgia for the perceived stability of the thirteenth century, a nostalgia that fuelled the demagoguery of John Ball and his followers in the Peasant's Revolt of 1381 (Zeeman, 2006). Pilgrims and pilgrimages are viewed by Langland through a cynical, critical lens. In The Prologue, Piers falls asleep and sees in his dreams the passing of great numbers of people.

The pilgrims he sees in the company of palmers, and they all claim they are in search of the pilgrimage sites of saints, such as Walsingham in Norfolk, England. However, Piers sees in their pious tales unholy lies – these pilgrims are travelling with women, they are work-shy idlers, and they are vain in their love of fashionable and distinctive capes.

Elsewhere, Langland describes a pilgrim met by Piers who wears on his hat badges and tokens bought as proof of his previous pilgrim visits: crosses, signs from Assisi, keys representing the Holy See of Rome, and sea shells that represent the Pilgrimage to the shrine of St James in Santiago (Webb, 2002). There is no doubt that Piers is suspicious of this pilgrim's motives. We are led to believe that this pilgrim, like those mentioned in The Prologue, is interested more in collecting the badges. Perhaps the pilgrim has not even visited all the places for which he has badges. The pilgrim is like the modern-day tourist who has to buy the tee-shirt and the fridge magnet, or the person at home who receives such things as presents – or indeed the liar who pretends that the pen that says, 'New York Giants' on it demonstrates that she has been to the states for a game, when in fact she has only been as far as an Internet shop.

Christian pilgrims made journeys to visit a host of local shrines, such as those at Walsingham. In England, of course, the shrine to Thomas Becket in Canterbury Cathedral made that holy site a must-visit for every faithful, free Christian in the country (Webb, 2007). Chaucer's (2003) late-fourteenth-century *The Canterbury Tales* gives us a detailed and satirical insight into the motives of travellers on the journey: some are seeking instrumental absolution for their sin-filled lives, others are glad to be away from their worldly troubles, others seek ways to make money out of fellow pilgrims and some are sincerely pious. There are rich and poor, men and women, on the pilgrimage – a relaxing of social barriers and prejudices. Chaucer's poem also gives us a rich account of the commercial activity that surrounds the pilgrimage, and its thick description of taverns, stables, peddlers and hawkers of souvenirs demonstrates the regular and lucrative nature of the pilgrim industry. The tales – told each night by a different pilgrim to his fellow travellers – represent the importance of story-telling as a leisure activity in a mainly oral culture (Mann, 1973). These tales also shed light on the everyday leisure life of people in England, with their stories of sexual intrigue, drunken escapades and public feasts (Boitani and Mann, 2004).

As well as local pilgrimages, mass pilgrimages on an international scale took place to Santiago, Rome and Jerusalem. To reach Santiago in north-west Spain, it was necessary to walk or ride hundreds of miles through southern France and northern Spain. Along the way, hostels were built that mirrored those built earlier in the Islamic East to serve the Muslim *hajj* to Mecca. Santiago and the towns along the way became rich servicing the free Christians of all classes who found their way along the trail. As mentioned, the seashell became the emblem of Saint James in Santiago, a favour to be bought and worn by those who had made the journey. Similar tokens were available to buy for pilgrims visiting Rome, who followed guidebooks written to help them find their way around the city's sites, and who stayed in buildings and streets dedicated to particular nations (Webb, 2002). As early as the ninth century, we can see evidence of this industry in the biography of King Alfred the Great, written by his secretary Asser (2004), where Asser describes how Alfred travelled to Rome as a young pilgrim and stayed with other Englishmen. Rome's survival as a city depended on Christian pilgrims visiting the Christian shrines and spending their money on secular pursuits. In this, Rome was similar to Mecca, which thrived on the religious compulsion to travel there: pilgrims needed to travel, eat and sleep in safety, and the Roman and Meccan merchant families provided what was needed for a price that was, for Christians in Rome at least, often extortionately high (Webb, 2002).

Further east, it was possible for Christians to make the long pilgrimage to Jerusalem, where a smaller, but no less commercial, proto-tourist industry thrived. A change in the local Muslim rulers brought restrictions and harassment of Christian travellers in the Holy Land, which resulted in the Crusades (Fletcher, 2004). While Outremer remained Christian, Jerusalem again became rich on the trade associated with Christian pilgrims. But as the Christians were beaten back – and despite the efforts of some nobles on both sides to protect the pilgrim trade – the capture of Acre by the Muslim armies in 1291 saw Christian pilgrimages to Jerusalem fall off completely. This, then, only made Santiago and Rome more attractive, as they were arduous (and hence good for the soul) but conveniently closer to home in lands populated entirely by Catholic Christians (Webb, 2002). The demand for seashell badges soared as hundreds more made their way to the Shrine of Saint James. The example of the seashell badge mentioned in Langland's *Piers Plowman*, then, allows us to see Christian pilgrimage in the Middle Ages as a nascent tourism industry, akin to the tourist industry associated with the Muslim *hajj* mentioned in the previous chapter.

Note

1. The Cornish game of hurling is described in a contemporary account, Carew's *Survey of Cornwall*, published in 1602 (cited in Strutt, 1969[1801], pp. 91–92). In it, Carew writes: 'Hurling taketh his denomination from throwing of the ball, and is of two sorts: in the east parts of Cornwall to goales, and in the west to the country. For hurling to goales there are fifteen, twenty or thirty players, more or less, chosen out on each side, who strip themselves to the slightest apparell and then join hands in ranke one against another; out of these rankes they match temselves by payres, one embracing another, and so passe away, every of which couple are especially to watch one another during play; after this they pitch two bushes in the ground, some eight or ten feet asunder, and directly against them, ten or twelve score paces off, other twain in like distance, which they term goales, where some indifferent person throweth up a ball, the which whosoever can catch and carry through his adversaries goale, hath wonne the game; but herein consisteth one of Hercules his labours, for he that is possessed of the ball, hath his contrary mate waiting at inches and assaying to lay hold of him, the other thrusteth him in the breast with his closed fist to keep him off, which they call butting... they must hurle man to man, and not two set upon one man at once. The hurler against the ball must not but nor handgast under the girdle, he who hath the ball must but only in the other's breast, and deale no fore ball, that is, he may not throw it to any of his mates standing nearer to the goale than himself.' In other words, no forward passes.

7
Leisure in Japan, China and India

The focus of this chapter is leisure in the history of Japan, China and India in pre-modernity and modernity. This chapter will avoid making Orientalist assumptions about civilization and the onset of Westernized cultures but will instead explore the relationships among political and social history, leisure, culture and globalization. The first section of this chapter will discuss examples of the traditional leisure form and practice in Japan, China and India: music and dance, tea-drinking rituals and chrysanthemum festivals (Siu, 1990). I will argue that traditions in all three countries need to be understood within the context of each country's individual histories. For each example of the traditional, I will then explore the circumstances that led to its establishment as a particular tradition, and the underlying tensions between the concept of leisure and the dominant philosophies and religions of each country. Set against this exploration of the traditional will be an account of historical change and the ways in which the meaning and purpose of these traditional leisure forms has been shaped by the tension between communicative agency and instrumental power. In the second section of this chapter, I will begin with an overview of modern leisure forms in each of these countries, followed by a more detailed account of particular examples in each country, such as the growth of popular music piracy in North India in the twentieth century (Manuel, 1993). I will argue that the instrumentality of twentieth-century capitalism, more than the instrumentality of state power, radically commodified and globalized leisure in these three countries – but at the end of the twentieth century and the beginning of the twenty-first there was still some communicative space to make choices within the constraints of that instrumentality.

Traditional leisure practices and pre-modern cultures

One of the most pernicious stereotypes of Orientalism is the idea that the East's traditional cultures were stagnating and in desperate need of the changes imposed on them by the colonizing West (Said, 1989, 1985). This stereotype, of course, is simplistic and inexact. While it is true to say that pre-modern cultures in some areas of the East were dominated by traditional or religious structures that emphasized order and continuity (Tu, 1996), every culture is a dynamic system (Geertz, 1973). Japan, China and India were places where such dynamism was expressed through tensions between and in different social classes, between conservatives and reformers, and between those who looked to the West for solutions to political crises and those who argued for a modernity infused with traditional notions of value. In this tension between the traditional and the modern, some scholars have preferred to use the less judgemental 'pre-modern' for the traditional (e.g., Reid, 2009). However, I will use 'traditional' as the term was used by modernizers and conservatives in each of these countries (in their native languages and in English), and such terminology is still used to understand the transitions, breaks and continuities associated with modernity in Japan, China and India. Furthermore, 'traditional' and 'modern' are useful terms to help us understand what leisure practices are described as such, and why they are described so.

India is portrayed by Orientalists as an unchanging tableau of caste obligations and servitude (Said, 1978), where a fatalist and primitive religion imbued in the masses and their rulers a strong belief in the cyclical nature of time. And if one believed that time was cyclical, then ideas of progress and change found it impossible to take root. It is no surprise that Western colonists and invaders perpetuated this stereotype of India and Hinduism – it suited the colonizers to portray the traditional, pre-modern cultures of India as outmoded, stifling and backward. This gave the colonizers an excuse to take control of people and territory in South Asia, a trend that saw the British take control over large parts of India, first through the late capitalist enterprise of the East India Company, then through direct imperial rule itself (Bowen, 2008). By the middle of the nineteenth century, following the mutinies of 1857, India was a major part of the British Empire, producing foodstuffs and minerals for Britain and sustaining the British fashions for curry and tea (Bayly, 1988).

In India, educated local reformers saw in modernity a way of embracing secularism and scientism. The advantages of Britishness – railways, bureaucracies, capitalism, cricket – were self-evident to one section of the local elites, mainly those living in British colonies and large cities (Darwin, 2009). Other local elites were fearful of the loss of Indian identity. Nationalist and conservative groups emerged to protect Indian identity and promote Indian culture and religions. Hinduism and Islam became strong markers of local identity (Bose and Jalal, 2003).

In the discussion over the nature of Indian culture, both modernizing nationalists and conservatives identified some welcome truths in the stereotype of the unchanging Oriental (Said, 1978; Bayly, 1988; Bose and Jalal, 2003). Continuities were observed and argued for in such practices as the caste system, Hindu ceremonies, gender relationships and leisure activities (Bose and Jalal, 2003). White British army officers, traders and scholars had already noted the unique scales and modalities of Indian music, and the preponderance of what they described as native folk dances (Bowen, 2008). Some British commentators argued that the music and dances were both religious rites and primitive releases of unbridled sexual urges (Ibid.). For local nationalists, the musical forms could be distinguished, using Western theories of aesthetics, between a classical form associated with courtly culture and the popular and localized rustic music played by lower castes for people to dance to. Just as Westerners formalized elite dance and music, so did Indian nationalist modernizers. They argued that classical Indian music and dance had a formal structure which survived unchanged from the Hindu royal courts, which preceded the arrival of the Europeans (Bakhle, 2005). For the conservatives, music and dance were identified with an authentic Hindu popular culture, which had been preserved unaltered since the beginning of the Hindu written record. Religious texts were found that described the dances and rituals of this earlier, mythical time of Hinduism (Mitter, 2001). In accepting this stereotype of unaltered authenticity, continuity and survival in their leisure practices, conservative Indian Hindus not only accepted Western notions of Orientalism, they also invented a tradition of purity (Bakhle, 2005). By stressing the Hindu nature of Indian music and dance, such leisure forms were opposed to the values and morals of Indian Muslims, who became the Other in early modern India (Van Der Veer, 1994). The music and dance also reaffirmed conservative, traditional gender roles and castes, ensuring the survival and reinforcement of such pre-modern notions (Bakhle, 2005).

In Japan, modernity came with the overthrow of the Shogun system and the Meiji restoration's commitment to Westernization. The early history of Japan is similar to that of China, typified by a tension between central, imperial control and local feudalism. In this political situation, a popular culture emerged that valued stability and order, with a distrust of mercantile endeavour and an honouring of masculine, military prowess (Cullen, 2003; Sugimoto, 2010). Under the shogunate, the samurai, traditionally the warrior caste which held land from the emperor in return for defending the land, took over the country completely in the seventeenth century (Totman, 2004). The emperor still existed, but the shoguns made the decisions and established the laws. They developed a social system that differentiated among four distinct castes of free men, with the merchants at the bottom of the scale (and women and minority ethnic groups below them) and the samurai at the top.

Inspired by traditional religion and Confucianism, the shoguns imposed order and stability across Japan, favouring and supporting traditions everywhere to the extent of discouraging any contact with foreigners (Cullen, 2003). At this time, for example, Christian missionaries and their local converts were crucified (Totman, 2004). Some historians have seen this period as a time of atrophy in Japan, both in terms of economy and of progress (Ibid.). Clearly, the system was damaging to the people who were not in the elite caste, and the system itself proved to be unsustainable due to the hereditary nature of the castes: the samurai who inherited their fathers' swords were no match for the armies of the modernizers in the civil wars of the late nineteenth century (Cullen, 2003). However, it is wrong to say the system preserved traditional, premodern Japanese culture without any change. In this period, the shoguns built roads to allow the samurai to move between their fiefs and the two centres of power: the city of the emperor and the city of the Shogun. These roads created a network for religious pilgrims to move to rural temples, and encouraged the growth of trade among different parts of Japan (Keene, 2006). This led to the growth of towns into cities, and the development of an urban culture that refined elite notions of art and propriety, as well as the dark leisure practices of theatres and inns (Hendry, 2003). The shogunate created a sense of nationhood and national culture out of a country otherwise divided into hard-to-reach regional fiefs (Keene, 2006).

Modernity came to Japan in the second half of the nineteenth century through a combination of nationalist feeling, pro-Imperialist sentiments among the new urban elites, and the imposition of relations with the West through the deployment of American gun-boats. The emperor established a modern constitution, eliminated the caste system and the samurai, and allowed some modern freedoms such as freedom of speech and assembly. There was, of course, a conservative reaction against these changes. Many of the old elite fought the new government's laws, but this defence of traditions against modernity was supported by many of the lower classes who saw in modernization the up-rooting of everything they took to be true: the natural and deified order of things, the distinctions of caste and the strict control of women (Cullen, 2003). This disaffection led to civil war and unrest, with much of the anti-modernization sentiment aimed at the Westerners and West-leaning elite who surrounded the new centres of political power in the public sphere: the ministries, the universities and the press (Jansen, 2002). This led to an increase in nationalism, and a belief that Japan could become a modern empire through taking some aspects of the West's modernity – factories, armies, navies – while retaining the devotion to the emperor and the natural order of pre-modern Japan (Hunter, 1989; Totman, 2004). These arguments held sway in the military, which had modelled itself on Western lines, adopting the professionalization, chains of command, technologies and physical training that gave Western empires such an advantage over other countries. This modernity was coupled with a religious dogma of Shintoism that recognized the divine nature of the emperor and the manifest destiny of the Japanese nation. In a cultural sense, the soldiers of the army were the inheritors of the samurai tradition, a belief encouraged by books at the beginning of the twentieth century that taught the modern soldiers the values of bushido, and sports such as judo and karate (Carr, 1993).

As Japan embraced its own kind of modernity, its new elites looked to the shogunate to find continuities in their leisure lives. Martial arts, linked to masculinity and the samurai, were obvious (re)inventions (Ibid.). Theatre was another place where Japanese nationality could be constructed, through the survival and (re)use of forms deemed to be classical in their nature (Jansen, 2002). In everyday leisure lives, the new urban middle classes looked to the West and to their own past (Totman, 2004). Some Japanese elites embraced the fashions and technologies of the West, especially following the First World War (Cullen, 2003). Others, however, looked back to the past, or a perception of the past, to find some stability and order in the ever changing world of modern Japan (Jansen, 2002).

The tea-drinking ritual was one domestic leisure activity invented anew for this modern age. In the shogunate and the imperial period before it, Confucianism had instilled a sense of decorum on every activity, no matter how great or small. Eating and drinking, among the urban classes and aristocrats of the shogunate, followed strict rules. This was an example of an Eliasian civilizing process at work to an extent (Elias, 1978, 1982) – more obviously, it was a display of instrumental power, showing the control the host or man of the house had over this domestic space, and displaying his wealth and good taste (what Bourdieu would call his cultural capital – see Bourdieu, 1986). Where the samurai set the fashion for this meditation and self-control in the dining room, the urban elites of the shogunate followed. For the modernizers of the Meiji Restoration, such rituals were examples of the stagnant torpor of the pre-modern past – part of the stereotype of the unchanging oriental (Jansen, 2002; Totman, 2004). Such disdain for tea-drinking rituals gave such rituals an attractive authenticity to those in the modern cities looking to find some connection and continuity to the unchanging world of a partly imagined past (Hendry, 2003). So, although the tea-drinking ritual was a leisure activity that survived into modernity from the pre-modern, it was the fact of modernity and its alienating power that made tea-drinking change its purpose to become a ritual of residual culture (Cullen, 2003). For the imperial nationalists, it had a limited function, as an expression of their sensibility. For the conservative elites, it allowed them to pretend they were a samurai, pampered by servants, living a life they had never led. For those caught between the past and modernity, tea-drinking was something comforting and familiar, even though the process was commodified through the industrialization of the raw products. Like beer-drinking in England, which changed completely in the grip of industrial capitalism, tea-drinking in modern Japan became a pastiche of its former self.

In the stereotyping of the East associated with Orientalism, China is portrayed as a mysterious and elusive country grounded in thousands of years of cultural stasis (Brook and Blue, 2002). Modern China is simply the same place as ancient China, and the leisure lives of the pre-moderns (along with the rest of their culture) are deemed to be essentially the same as the modern Chinese. This, of course, is a gross simplification of the complex history of ancient and modern China. Traditional, pre-modern China went through a series of profound cultural and political changes (Fairbank and Goldman, 2006). However, despite the cycle of centralization and breakdown, and the invasions of Mongols and Manchurians, there was a continuity of philosophical and religious beliefs in both high and popular cultures in traditional China.

The unifying factor in Chinese identity was Confucianism, the writings of Confucius that were preserved and used from their creation in the sixth and fifth centuries BCE through to the twentieth century (Clements, 2004). Confucianism provided a guide to ethics, politics, domestic life, medicine, public health and administration. It drew on and expanded earlier Chinese philosophies such as Taoism, and for its first thousand years was the subject of much refinement and commentary by other elite philosophers (Yao, 2000). Confucianism put a premium on social order in this world, which ideally should reflect the perfect order of the higher worlds of the heavens. It was the duty of every individual to know their place in the social order, to accept the rules of behaviour associated with that place, and to ostracize those who went against this 'natural' way of things. So the peasant tilled the field, the shepherd tended flocks, the mandarin collected taxes and the emperor reigned over the entire world (Yu-Lan, 1997).

Confucianism, in one sense, was clearly a way of maintaining existing political power through the legitimization of inequalities in traditional Chinese societies: men were superior to women, and higher castes were superior to lower castes and foreigners. This justified the reign of the elite through the use of the instrumental rationality of Confucianism. It appeared to be a rational and sensible way of being, which allowed philosophers to make many breakthroughs in natural science (Lloyd and Sivin, 2004). For the majority of Chinese people, there was no other way of thinking about their situation and the hegemonic control under which the elites kept them in control. Rebellions, of course, did occur – driven by famine, when the imperial systems failed, or when the empire fractured into dozens of petty kingdoms that were unable to maintain that hegemonic control – but these were exceptions to the rule of Confucian order. When foreign invaders from the north took control of pre-modern China, they quickly adopted the Confucian way of ruling and justifying that rule (Yu-Lan, 1997). When Western countries started to win concessions for trading posts in China, and the nineteenth century saw those Westerners win power from the local elites, then Confucianism became disrupted. Resistance to Western interventions came from modernizers, nationalists, socialists, warlords and religious fanatics (Cohen, 2003).

Confucianism allowed space for leisure practices. For the elites of pre-modern China, leisure sites divided into the formal rituals of the court and temple, and the informal and private domestic activities of the home. In both sites, men and women were expected to conform to their gender roles, and to respect the hierarchies of class and caste. Sports and physical activities were seen as not befitting proper, elite gentlemen. Instead, leisure for these elites involved contemplation, reading, and listening to recitations and music. In the public spaces of the court and temple, men could become involved in strenuous activities such as archery and hunting, but these were linked to military training and instrumental obeisance and were not free choices made for the love of physicality. At home, men were expected to continue to behave with the same level of decorum and gentility, and domestic leisure activities were a mirror (in a sensibly Confucian way) of the public leisure rituals: the family obeyed the hierarchies of the social order, the man read and contemplated, or ate his food slowly, while the women and servants waited upon him (Ko, 2003).

For the men of the lower classes, there were fewer constraints on behaviour, and therefore probably more communicative choice within the limits of their leisure time, their money and the laws and ethics of the land (Fairbank and Goldman, 2006). There were, in towns and cities, opportunities to visit taverns and brothels, to gamble, and to watch theatre shows. In the countryside, there were chances to walk, fish and hunt, away from the prying eyes of officials and landlords (Rowe, 2001). Away from these moments of communicative action, leisure for the masses was viewed through the instrumental lens of state control and Confucian social ethics. Work defined everybody's status and place, and for most people leisure time was brief and of the domestic space. Where time away from work was available, it was filled with festivals that marked the cycle of time, the passing of seasons and the worship of the various totems, ancestors, heroes and gods that filled the Chinese religious landscape.

Like the holy days of Western Europe in the Middle Ages, these were tightly controlled and policed ceremonies, with strict rules about what was acceptable behaviour (Fairbank and Goldman, 2006). Some dissenting voices and attitudes were allowed, but only as part of the rituals themselves – so, for instance, women and lower classes were allowed some autonomy to take part in some way, by, for instance, cheering on processions from the side of the street, or lighting fireworks of their own (Cohen, 2003). Some people naturally saw festivals as an opportunity to put their tools down and to meet their friends, chat, tell jokes, flirt and so on (Rowe, 2001; Fairbank and Goldman, 2006).

Chrysanthemum festivals were an example of these Confucian rituals (Siu, 1990). At first, these were associated with more general spring festivals but became a way of identifying with a conservative, traditional Confucian social order. In the confusion and trauma of the Western intervention in China, such festivals were vital leisure sites for preserving the social world and reaffirming its relationship with the perfect worlds of the heavens (Yu-Lan, 1997; Yao, 2000). By the twentieth century, the conservative reaction against modernity meant the festivals, like the temples and the private rituals, became re-invented. In a modernizing nation torn apart by factions, dispossessed of the imperial head of state, Confucianism and its leisure traditions were perceived to be the essential remnant of a purer, more ideal world, which by the middle of the twentieth century was perceived to have vanished.

Leisure in modern Japan

In the twentieth century, Japan was subject to two modernizing trends. The first, the modernism associated with the imperial nationalism of the new nation-state, was (almost) consumed in the bombings of Hiroshima and Nagasaki. The Japanese nation-state had developed a strong industrial base and advances in technological capitalism by the 1930s – and there was some evidence among the elites of changing societal attitudes towards women, and the adoption of Western fashions by elite, educated and urban women (Harootunian, 2001). Industrial growth and rapid urbanization saw the creation of what we would describe as modern leisure industries: popular literature and music, cinema and dancing. However, this superficial Western modernity hid the deep roots of tradition and conservatism wrapped up in Japanese nationalism. In this pre-war Japan, modernity was associated strongly with the expansion of the empire and military victories. After the defeat in 1945, Japan faced the second of the modernizing trends, which was imposed on it by the American occupying forces: a pacifist constitution, a shift from nationalism to liberalism, and an embrace of globalization, including free trade (Allison, 2004). Both these trends shaped modern Japanese society through the second-half of the last century and into our own. On the one hand, Japan was admitted into the company of Western liberal democracies, engaging in international trade, modern science and gaining a reputation for technological advances in telecommunications, motoring and engineering, and computing (Gordon, 1998). This led to the development of a distinctly Western modern identity, one that saw Japan's post-war generation embrace individualism, Western tastes such as rock music and baseball. On the other hand, modernity was framed by the facts of Japanese life: the emperor still reigned (and still reigns today), the nationalist traditions such as the Shinto religion and the temples to the war dead survived, and many writers lamented the loss of the invented traditions of bushido and the samurai way just as popular culture took on a distinctly American sheen.

One of the ways in which these struggles over the meaning of Japanese identity and society were fought was in the sphere of leisure and culture. The economic success of Japan from the 1960s to the early 1990s saw a huge growth in material goods and wealth in the average Japanese home. The post-war generations grew up with a liberal education, access to income and a desire to express their identity through the consumption of various modern fashions (Iwabuchi, 2003). Young women were able to find the freedom to play at constructing femininities that challenged the traditional domestic roles of their grandmothers (Martinez, 1998). What started as a movement in the middle classes was soon replicated by Japanese women and girls across all classes: the availability of consumer goods, role models and, latterly, the Internet, all have allowed Japanese to play with a number of (post)modern, liquid identities, from post-punks to emos. All these identities provide a sense of community and belonging that embrace some elements of the traditional past (spirits and magic, the make-up of the geisha and theatre) while taking, magpie-like, from a hundred other sources to create a myriad of hyphenated subcultures. The related Japanese taste for manga cartoons has influenced these fashions and neo-tribes to such an extent that a word has been created – 'cosplay' – to capture the fashion of dressing up as characters from these cartoons (and films and video games – see Brown, 2008). Such female expressionism and communicative action is, of course, viewed as a social problem by some commentators in modern Japan. For those from the conservative right, seeing Japanese girls dressed as heavily made-up manga dolls and drinking is an affront to the traditional values of the male public sphere (Napier, 2008). For some of those on the left, the cosplay is seen as a commodification of female sexuality, one that makes young women the subject of a male gaze (Martinez, 1998): rather than empowering young women, the fashions stupefy and objectify them.

Another leisure pastime in modern Japan is gaming (Allison, 2004). Initially, Japanese teenagers bought into the concept of role-playing games in the 1970s (Hjorth, 2011). With the advent of computer technology, Japanese companies became world leaders in the digital gaming industry, inventing a string of video games and consoles that became well-known at home in Japan. Soon computer games were seemingly in every home in Japan. Furthermore, games drew upon traditional, pre-modern Japan as well as the fantasy and science-fiction genres, taking concepts about caste from the shogunate: the samurai as the exemplar of the heroic and virtuous fighter, the untrustworthy thieves of the towns, and the court finery of the emperor. With the spread of gaming, and the growth of the

Internet, came more concerns about young people's leisure time. Young Japanese men, in particular, had already been identified in the 1960s by conservatives as being susceptible to introversion and obsession with their toys, books or television programmes (Goto-Jones, 2009; Sugimoto, 2009). The gaming industry seemed to magnify this trend. Young men and boys could now spend many hours every day playing their favourite games, on a hand-held console or in a multi-player game on-line. In both cases, the addictive nature of the games mean it is very difficult not to keep trying to get to the next level, or to find the next piece of treasure. Again, this anti-social leisure phenomenon has been the subject of criticism from both conservatives and intellectuals, the former believing that games spoil the natural spirit of boys, the latter arguing that such games are tools of instrumental control and stupefaction.

Finally, sports are a key site for defining identity in modern Japan. The post-war generation adopted baseball because of its obvious relationship with American popular culture, and this sport remains popular among participants and paying fans (Sugimoto, 2009). However, other sports are as equally important in various ways to different parts of modern Japan. Football has become a new way of expressing a liberal, Japanese identity that comes unadorned with Americanism (Goto-Jones, 2009), and the World Cup of 2002 proved that professional football has a strong fan-base. But for those modern Japanese who still associated with the imperial throne and the traditions of the pre-modern, the martial arts and sumo remain as (invented) continuities into that masculine world of status and honour (Reader, 1989; Saeki, 1994).

Leisure in modern China

In the second half of the nineteenth century, China started to feel the twin pressures of modernity and nationalism. For many of China's educated, urban classes, the empire's failures – especially those political losses to the Western powers – were an indictment of traditional, Confucian values (Pomeranz, 2000; Mitter, 2004; Fairbank and Goldman, 2006). Some people looked to the example of Japan, in its own modernizing throes, for evidence of the need for social, political and economic change. Others agitated for a radical shift in social power, a redistribution of wealth alongside the establishment of a republic. Conservatives attempted to maintain the traditional social order, but by the early twentieth century a new class of entrepreneurs, industrialists and western-educated intellectuals led a coup against the Imperial throne. The emperor was overthrown, and the mandarin system abolished: replaced by a national republican constitution and western-style schools and universities. A period of civil unrest followed, in which the coup leaders were deposed, and large parts of the country fell under the control of warlords. In the capital city, Chiang Kai-Shek came to power promoting Chinese nationalism in a thoroughly modern sense – reusing some of China's Confucian culture, while campaigning vigorously for physical education, women's liberation and free trade (Mitter, 2004). In accepting Western support, the new Nationalist regime of the Kuomintang distanced itself from more radical reformers. The Communist Party was an obvious site of opposition to the Nationalists, though they had much in common, too. The Communists were also modernizers, and appealed to Chinese nationalism and commitments to abolishing some of the inequalities of pre-modern China. But the Communists wanted to go much further than the Nationalists in tackling the landlords and capitalists who continued to make capital from the labour of the rural and urban poor (Pomeranz, 2000).

When Japan invaded Manchuria in 1931, the Communists and Nationalists made common cause in trying to beat back the invaders. But there was no love between the two modernizing parties of the country, and both struggled to gain the upper hand over the other – the Nationalists backed by the United States, and the Communists by the Soviet Union. By the end of the Second World War, the Nationalists were still ensconced in Beijing, and still recognized by the post-war settlements as the legitimate government of the nation (Mitter, 2004). But the Communists were gaining more support in the great swathe of countryside in the heartlands of China. Promises of equality were balanced by security and the availability of food. The Communists soon pushed onto Beijing, forcing the Nationalists to flee to Taiwan in 1949 (where they set up a Chinese Government of their own, in defiance of the mainland and the native Taiwanese).

Communist China inherited the modernizing programme of the Nationalists, though they discarded the reliance on the West and the support for the industrial capitalist class. Under Mao Tse-Tung, China's economy was developed following the state capitalist model of the Soviet Union, with a uniquely Chinese focus on the peasant class of the countryside as an ideal role model (Pomeranz, 2000). Confucianism was inverted, and in the second half of the twentieth-century China went through a number of top-down programmes, which eliminated the landlord classes, nationalized industries and ultimately led to the Cultural Revolution, where everything associated with the West and the intellectual classes was deemed un-Chinese.

In the period of the Nationalists, a modern, urban middle class had emerged in Beijing and Shanghai, which adopted Western fashions in culture and leisure. Some intellectuals tried to preserve or reproduce aspects of traditional Chinese culture, such as the theatre and opera (Mitter, 2004), but for many of the new urban elite, educated in the ways of the West, leisure time was spent playing at being a Westerner: dancing to jazz, smoking cigarettes, watching films, drinking cocktails (Fung, 2000). It was this Western influence that was challenged by the Cultural Revolution. Under the reign of the Red Guards, any leisure life-styles associated with the wealthy, Westernized elites were deemed to be depraved (Clark, 2008). The Cultural Revolution went further from merely banning things that were judged decadent by instilling in the Chinese masses a commitment to use their leisure time in a controlled and productive way. Nationalist festivals and ceremonies were (re)created for mass participation and instrumental conformity (Mitter, 2004), some traditional Chinese plays were judged to be educational because of their correct themes (Clark, 2008), and Chinese youth were encouraged to take part in sports and physical recreation, such as gymnastics and physical training (Riordan and Jones, 1999).

Inevitably, the Cultural Revolution produced a backlash, brought about when the Chinese economy collapsed (Pomeranz, 2000). At first, the ruling elite of the party was torn between defenders of the aims of the Cultural Revolution and those who saw a possible third way of economic liberalism. The pragmatic modernizers won the battles inside the Party, and a period of economic liberalism followed in the late 1970s and 1980s. This saw more individual freedoms, too: students were encouraged to study at universities in the West; company executives travelled the globe to deal with clients; and the new generation of party cadres had money and passports to allow them to visit the West as generous tourists (Barme, 1999). For those who could not visit the West themselves, the new capitalism brought a wave of Western products, fashions and leisure trends. Football and other professional sports became popular across the country (Stockman, 2000). Western pop music and films were pirated by Chinese firms and sold alongside their Chinese equivalents, which were influenced by the fashions of the West (Hutchings, 2000). Western television programmes were broadcast on Chinese TV, and local stations and producers started emulating the demand for home-grown soap operas, detectives, and reality TV programmes (Mitter, 2008). In the cities of China, a Westernized generation of middle classes grew up on fast food and video games. In the poorer areas of the countryside, however, ready wealth and free time were not easy to find for such leisure (Stockman, 2000).

By 1989, the Chinese system had opened up to such an extent that some intellectuals demanded the logical end product of communicative choice: democratic freedoms and an end to arbitrary repression and unaccountable, instrumental hegemony. Activists gathered in the centre of Beijing to call for the political freedoms that must accompany economic freedoms if communicative action is to be possible. The activists were crushed violently, and the survivors were imprisoned or exiled. Since that time, China has continued to follow a Westernized, modern programme compete with economic freedoms, but the Communist has retained its grip on political freedom, using the excuse of Confucianism and the uniqueness of China's traditions to suggest that political freedoms are somehow 'un-Chinese'. When Chinese TV bought the rights to the Pop Idol programme and launched its search for a new pop star in 2005, the viewers casting their votes were the nearest the Chinese have got to exercising their democratic right. But the Government was so wary of even this, that they changed the words of the format, so that the vote was called something far less binding or communicative: they were merely texting in messages of support for their favoured singer (Mitter, 2008).

Modern India: Bollywood, tape-trading and cricket

Modern India was born out of the tumult of the independence struggle, and the religious strife that saw India and Pakistan created out of the British colony in 1947. India established itself as a democratic, secular state – in contrast to the dictatorship of China, or the Shinto-aligned constitutional monarchy of Japan. However, its secular nature fit uneasily with the massacres and translocations of populations in 1947, when Muslims fled to Pakistan, and Hindus to India, before the new states were created. The secular spirit of India's modernity legislated against caste discrimination, religious discrimination, and even gender discrimination (Bose and Jalal, 2003). However, in practice, much of India remained Hindu in religion and culture, and Hinduism has dominated high culture, popular culture and leisure in India since independence.

Modern India's most famous cultural export is the films of Bollywood. Watching films and talking about the celebrities is a leisure pastime of millions of Indians of all classes, castes, religions and genders (Mishra, 2002). It is a phenomenon that has created millions of pounds for the industry, the key producers, actors and music composers such as A. R. Rahman (Dudrah, 2006). Bollywood films have spread out of their original Indian home into the neighbouring South Asian states of Pakistan, Sri Lanka and Bangladesh, and even as far north as Afghanistan (Bose and Jalal, 2003). They have also followed the post-colonial, diasporic movement of Indians into the West (Kavoori and Punathambekar, 2008). The films are spectacles, infused with Hindu traditions and values (sometimes drawing explicitly on Hindu sacred texts), but with distinctly modern tones of individual love and romance (Dudrah, 2006). The films are musicals, recycling Indian classical and pop music, drawing on Western pop influences, and underpinning extravagant dance scenes that combine traditional steps with modern, American dance-forms (Mishra, 2002).

The popularity of Bollywood films, combined with advances in video technology in the 1980s, led to the development of a huge black market in copied tapes. Some tapes were traded through postal and informal networks, shared between contacts, and others were sold at markets and by transient traders. These were not just Bollywood films, but also Hollywood films, music cassettes and pornography. The illegal trading and copying extended into Pakistan and Afghanistan, places where censors were more likely to refuse permission for a film's release on moral grounds, or where getting legal permissions in tangled bureaucracies was difficult. Tape-trading occurred through the globe in the 1980s and 1990s, before the rise of the Internet and on-line file-sharing sites – it was particularly strong in alternative music scenes (Bennett, 2001). However, the sheer size of the South Asian economy, and the increasing demand for watching films and listening to music as a communicative leisure choice, saw the expansion of this black market. In effect, the commodification of Bollywood's products encouraged consumers to find ways of consuming those products that did not contribute to the industry itself (Dudrah, 2006). Such consumption, of course, was a deliberate communicative choice in times and places when the romantic element of the films was questioned by arbiters of morality (Mishra, 2002).

Another important leisure form in modern India (and Pakistan) is cricket. As an English cultural icon, cricket is more than a game; it signifies something overarching, religious, almost totemic (Searle, 2001). For Williams (1999), it is an institution expressing a distinctively English set of ideologies and subsequently plays a significant role in how the English imagine themselves. The coalescence of cricket and England was very much a creation of the Victorians (Edensor, 2002). Unquestionably, cricket emanates powerful visions of an unspoiled rural England – the image of a village green is particular evocative.

Cricket has also come to represent a microcosm of England's imperial rule and thus continues to pull on the heart strings of the nostalgic

English punter (Williams, 1999). Norman Tebbit's infamous 'cricket test' exemplifies the importance of cricket for English culture's unchallenged continuity (Wagg, 2004, 2007). Similarly, James (2005[1963]) noted how cricket occupied a central site in many of the anti-colonial struggles between colonizer and colonized. Cricket is an English creation and encapsulates many of the characteristics needed to be a stereotypically Victorian and Edwardian Englishman, namely: soul, temperament, strategy, diligence, hard work, ruthlessness, pride, respect, 'manliness' and, more often than not, White middle class (Rutherford, 1997; Marqusee, 1999).

So often cricket is held up as the archetypal elitist institution; run by 'toffs' for toffs. No aspect of the game is more responsible for this than the gentleman-amateur dichotomy which frequently caricatures the game. For over two centuries, the distinction between playing status epitomized the classed nature of cricket and moreover, accentuated perceptions of cricket's exclusivity. To this day, cricket has strict symbolic boundaries. For all its democratic utterances, cricket remained the domain of the English elite. Importantly, this was not just in terms of administration and spectatorship. Cricket was the first team sport and, moreover, the first sport played on foot, in which the upper classes participated. Principally, however, the agenda of the ruling classes and those occupying positions within the 'powerhouse' of world cricket – the Marylebone Cricket Club (MCC) – was to successfully export the laws of the game; first to the colonies and the rest of the world thereafter. Indeed, the aim and purpose of the MCC sending teams all around the world was to 'spread the gospel of British fair play' (Marqusee, 1994, p. 93). Hence, a set of English laws would accompany the team, and matches would be played with the desired etiquette and civilization of the Western world. For Baldwin (1926, in Williams, 1998), cricket had a responsibility for communicating English moral worth with 'races' less 'civilized' than *our* own. He further argued that English imperialism was functional; aiding the development of the indigenous peoples it touched:

> [S]preading throughout such parts of the world as we control, or in which we have influence, of all those ideas of law, order and justice which we believe to be peculiar to our own race. It is to help people who belong to a backward civilisation, wisely to raise them in the scale of civilisation. (cited in Williams, 1998, p. 102).

Hence, cricket is not just a constituent of an 'invented tradition' (Hobsbawn and Ranger, 1983) of English national identity, it helped create it (Birley, 2003). Disguised by utterances of social inclusion was a very specific agenda of cultural homogenization, believed to be possible by the apparent rigidity of ethnocentric imperialist hierarchy (Crabbe and Wagg, 2000). Indeed, international cricket originally grew out of the modernist, rationalized sports projects associated with the British Empire (Crabbe and Wagg, 2000; Birley, 2003); an empire which has dwindled into insignificance as British colonies have been granted independence (Holt, 1989).

Cricket's symbiosis with the past, its status as an English national relic (Marqusee, 1994) and its symbolism are not accidental or incidental. Cricket boasts a longer history than other sports, but there is more to it than that. Cricket has evolved into what it is because of its birth in a particular place and time. Of course, this is the same for every sport, but for cricket it is especially true. Not only has the game globalized, there also has been the globalization of an ideology; of local and national myths and its players, and moreover, the commodification of an imagined national identity.

For all the game's Englishness, cricket has lost its cultural centre of gravity to the South Asian subcontinent. In colonial times, the English rulers reluctantly allowed the locals to play cricket – it spread in spite of the British Empire, through the enthusiasm of missionaries, schoolmasters and Anglo-Indians, as well as the passion of a few elite Indians educated in England. It was soon adopted by Indian men and boys en masse through schools and those two engines of imperial progress: the army and the railways (Bose and Jalal, 2003). Cricket was seen as something owned by Indians, or at the most, grudgingly shared with the English. For the best cricketers, high status and material rewards were available, through the system of tours, playing for clubs in English leagues, or finding a job in an office with good pay a world away from the harsh working lives of the average Indian man.

With the decline of amateurism in the second half of the twentieth century, Indians embraced professional cricket as a path to proving their masculine worth. Indian cricketers became a familiar site in English league cricket, exotic players paid to play to entertain white, working-class Lancastrians and Yorkshiremen (Birley, 2003). Professional leagues in England and Australia became more lucrative with the commercialization of elite cricket in those countries, brought about through the selling of television rights and sponsorship. Back in India, spectators went to test matches in their thousands, but millions more followed the games on television, on the radio, in newspapers and in drinking dens and gambling syndicates.

In India, the Premier League has generated immense levels of popularity; made fortunes for players, book-makers and owners; and demonstrated to the world that India is the home of modern, globalized and commodified cricket (Mehta, Gemmell and Malcolm, 2009). In the Indian Premier League (IPL), though, can be seen something uniquely Indian: the decorum of the cheerleaders combing the morals of Hinduism with the objectification of the Western male gaze; the nicknames copying the rule-book of sports marketing first written by the NFL in America, but derived from Indian and Hindu mythology; and the billionaire owners from the uniquely Indian world of Bollywood. Even where those owners make their living from something that might be seen to be Western, there is an Indian twist. The 'Bangalore Royal Challengers' were bought at auction by Vijay Mallya, one of the richest whisky barons of India, and owner of a number of distilleries in Scotland. Whisky is big business in India, and seen as something urban, modern and cosmopolitan, drunk by men and women alike who emulate their Bollywood idols who sell the products in flash adverts (MacLean, 2010).

Conclusions

Even though capitalism in China was regulated by the state, it is obvious that the instrumentality of twentieth-century capitalism, more than the instrumentality of state power, radically commodified and globalized leisure in these three countries. The global politics of economic progress, first advocated by Japan, has been picked up by India and China, who seem intent in the twenty-first century to compete with each other for global dominance in the marketplace. This has led to Japan, China and India all encouraging the commercialization of leisure, which in turn has contributed to instrumental leisure forms becoming dominant in each country: the passive consumption of sports and music, and the packaged tourist trips that already made places like Howarth in West Yorkshire, England, put up Japanese language signposts to Top Withens for the benefit of middle-class tourist seeking out the landscape of Emily Bronte's *Wuthering Heights*. But as I have shown, at the end of the twentieth century and into this one, there was still some communicative space to make choices within the constraints of that instrumentality. In Japan especially, where such communicative freedom has led to the flowering of many subcultures, this is clearly happening. In India, the gap between rich and poor still restricts the masses from participating in the (post)modern wish-making of whisky cocktails. In China, while political freedoms remain restricted, it seems the only communicative leisure spaces and forms may exist in countercultural states, where music and literature are used as dissent.

8
Early Modern Leisure

This chapter returns to Europe and focuses on the Early Modern period: with a particular focus on the development of a politicized, popular culture associated with the growth of agency in the public sphere (Burke, 1999). The first section of this chapter will explore the wider context of the rise of humanism as a reaction to the power of the Catholic Church. I will argue that competing notions of freedom, culture and morality challenged medieval Christian views of leisure. I will examine here the Reformation and the retreat from work into rural spaces by some Protestant groups that stemmed from communicative desires to find authenticity in nature. I will claim that this desire was an echo of the later Romantic Movement of the nineteenth century, which in turn foreshadowed the rise of environmental tourism in the twentieth. At the same time, rulers and states started to recognize the instrumental uses of leisure. The second section of this chapter will examine the rise of bourgeois culture in the free cities and states of mainland Europe, using the example of Venetian festivals and popular culture to argue that leisure increasingly marked out a site where hegemonic powers used instrumentality to construct political cohesion. In the third section of this chapter, I will discuss leisure in the work of Shakespeare and how his account of leisure and the history of leisure need to be understood in the context of the construction of Englishness and Anglicanism. The final section of this chapter will discuss the religious and social struggles of the 1600s, concentrating specifically on the example of the rise of Puritanism in England, its transferral to the New World, and the English Civil Wars. I will argue that during the Civil Wars and afterwards, leisure became the site of an instrumental struggle over morality and meaning, between reformers seeing the work of the Devil idle hands, and Monarchists who used the suppression of popular culture and leisure as means to gain support for their anti-Puritan campaigns.

The Reformation and free thinking

The Reformation emerged from medieval concerns with the excesses of the Church (such as the selling of indulgences satirized in Chaucer), and Renaissance concerns with purity and returning to 'true' wisdom (Lindberg, 2009). Initially, when Luther (allegedly) nailed his 95 theses to the door of the church in Wittenberg, there was hope that his complaints would be addressed, and a reformation undertaken of the Church's policies (Ibid.). Many princes, people and priests in Northern Europe agreed with the key messages of Luther's theses: that the Church needed to follow the example of Christ and not covet extreme wealth and hegemonic political power. The proposals for reformation were, however, rejected by Rome, and the protesting German princes gave their name to the Protestant movement (Ibid.). Alongside conservatives reluctant to break with the Catholic Church entirely – and those careful to ensure the medieval justification of feudal power as a temporal enactment of sacred Law – there quickly arose a torrent of 'puritan' and messianical movements. The former tried to return to an imagined purer version of Christianity, untainted by the accretions of medieval Catholicism; the latter found in the freedom of interpretation of sacred texts the impetus to make predictions about the return of Christ (Ibid.). Both of these movements were the result of the free thinking legitimized by the Reformation, and in turn both provided fertile ground for more free thinking to emerge. This is the point in history when having leisure time to think, and the freedom to think for oneself, established a Habermasian communicative sphere.

The career of Paracelsus is instructive in helping us understand the impact of the Reformation on free thinking and, in turn, on leisure. Paracelsus flourished from around 1520 until his death in 1541, in which time he wrote a great number of works pertaining to medicine, theology and alchemy. In his life, Paracelsus was caught in the religious turmoil of the Reformation, and Weeks (1997) has shown how the varied and often contradicting corpus of Paracelsus' work can be seen as a theological enterprise placing revealed knowledge of the Divine in nature, an attempt by Paracelsus to complement and criticize the reforming theology of Luther, to whom he was often compared. Yates has argued that Paracelsus was informed by the neo-Platonic writings of Ficino, and has identified hermetic influences in the Paracelsian doctrine of signatures

(Yates, 1964). But Paracelsus was also undoubtedly in debt to earlier traditions of magic such as the alchemical work of Albertus Magnus and Johannes Trithemius (Kieckhefer, 1989).

Key to our understanding of Paracelsus and his influence on the debates around the philosophy of nature is his belief that the human body is a mirror of the cosmos at large (Henry, 1990). This mystical and alchemical relationship between the microcosm and the macrocosm, between the Trinity and the three alchemical elements of sulphur, salt and mercury, gave rise to the doctrine of signatures, a semiotic analysis of nature as a revealed text. Paracelsus believed that Divine virtue had created resemblances that could be discerned and understood in nature by those philosophers who were prepared to read them. And this preparation was empirically based. If the signs of the divine were placed in nature, then the philosopher had to examine nature to understand the knowledge those signs revealed. In other words, Paracelsus advocated an active, scholarly leisure life, seeking out the signs of the Divine in the world around him. Paracelsus, then, argued for the freedom of rational humans to find knowledge for themselves, a typically Puritan and Protestant way of thinking. Such freedom of thought inevitably encouraged freedoms in leisure – using one's free time to find things out, to explore the countryside, to write poetry, to find the comfort of alcohol and good company (Weeks, 1997).

Paracelsus had a radical agenda. In Basel, he burned the works of Avicenna and dismissed Galenic medicine and its humoural basis. He lectured and wrote in German. He contributed to the eschatology of the apocalypse (Webster, 1982). But he inspired many followers in alchemy and medicine. By the 1580s, long after the man who encouraged his followers to defecate on received, scholastic sources had died, Paracelsian medics were challenging Galenic physicians for the favour of courtly patronage (Debus, 1978). The Paracelsians, such as Severinus at the court of Copenhagen in the 1610s, combined the popular appeal of Paracelsianism with this support from influential patrons, to introduce chemical practices and remedies to the field of medicine. And although the Thirty Years War swept away the more utopian ideals of these courts, Paracelsianism survived as an influence on the development of medicine and chemistry in England (Yates, 1972; Debus, 1974).

Paracelsus' instructions to his followers to find signs of the Divine in nature ('as above, so below') echoed the words of earlier humanists and radicals of the Renaissance, such as Ficino and Pico della Mirandola (Henry, 1990). The inspiration for these authors were the 'Ancient Egyptian' texts of Hermes Trismegistus, which, although later proven to be fakes, strongly influenced Renaissance views of nature and the secrets of nature (*Third Book of The Holy Divine Pymander* by Trismegitsus, from the 1650 translation by John Everard, found online at http://www.sacred-texts.com/eso/pym/pym04.htm):

1. The glory of all things, God, and that which is Divine, and the Divine Nature, the beginning of things that are.
2. God, and the Mind, and Nature, and Matter, and Operation or Working, and Necessity, and Matter, and Operation or Working, and Necessity, and the End, and Renovation.
3. For there were in the *Chaos* an infinite darkness in the Abyss or bottomless Depth, and Water, and a subtle in Spirit intelligible in Power; and there went out the Holy Light, and the Elements were coagulated from the Sand out of the moist substance.
4. And all the Gods distinguished the Nature full of Seeds.
5. And when all things were interminated and unmade up, the light things were divided on high. And the heavy things were founded upon the moist Sand, all things being Terminated or Divided by Fire, and being sustained or hung up by the Spirit, they were so carried, and the Heaven was seen in *Seven Circles*.
6. And the Gods were seen in their *Ideas* of the Stars, with all their signs, and the Stars were numbered with the Gods in them. And the Sphere was all lined with *Air*, carried about in a circular motion by the Spirit of God.
7. And every God, by his internal power, did that which was commanded him; and there were made four-footed things, and creeping things, and such as live in the water, and such as fly, and every fruitful seed, and Grass, and the Flowers of all Greens, all which had sowed in themselves the Seeds of Regeneration.
8. As also the Generations of Men, to the Knowledge of the Divine Works, and a lively or working Testimony of Nature, and a multitude of men, and the dominion of all things under Heaven, and the Knowledge of good things, and to be increased in increasing, and multiplied in multitude.
9. And every Soul in Flesh, by the wonderful working of the Gods in the Circles, to the beholding of Heaven, the Gods Divine Works, and the operations of Nature; and for signs of good things, and the Knowledge of the Divine Power, and to find out every cunning Workmanship of good things.
10. So it beginneth to live in them, and to be wise according to the operation of the course of the circular Gods; and to be resolved into

that which shall be great Monuments and Rememberances of the cunning Works done upon earth, leaving them to be read by the darkness of times.

11. And every Generation of living Flesh, of Fruit, Seed, and all Handicrafts, though they be lost, must of necessity be renewed by the renovation of the Gods, and of the Nature of a Circle, moving in number; for it is a Divine thing that every worldly temperature should be renewed by Nature; for in that which is Divine is Nature also established.

These hermetic-influenced fifteenth and sixteenth-century humanist writers called for Europeans to return to a wilder, pristine rural state, or at least for Europeans to recognize the beauty and value of nature. They influenced later utopian thinkers such as More and Campanella, establishing the opinion that how society is ordered on Earth should reflect the Divine (Yates, 1972; Debus, 1978; Webster, 1982) – including how we use our free time and leisure. Both the instrumental use of the countryside – its trees, its herbs and fungi – and its communicative role – its aesthetic value, its tranquillity – were advocated (Yates, 1972).

For the ruling classes, the turn to the Divine in nature was expressed through the leisure spaces of formal gardens. The Renaissance and Early Modern periods saw the invention and emergence of the formal European aristocratic garden (Mukerji, 1994). These gardens were laid out according to mathematical rules taken from Classical and Neo-Platonic sources, and planted with herbs, grasses, flowers and bushes that provide aesthetic, spiritual and healing properties. These leisure gardens were the preserve of the elite, with inner sanctuaries for private play and more formalized, outer places for public entertainment (Borsay, 2005). Every Renaissance prince had to have a garden better than his rivals, conforming to the latest fashions and philosophies, and allowing for the latest leisure trend, be it mechanical musicians, or tennis (Mukerji, 1994). As Jardine (2009) explains, the formal garden was something with which William of Orange was obsessed – when he invaded England in 1688 and marched into London a triumphant Protestant king, he rode away from the cheering crowds just so he could see for himself the gardens laid out for the Stuarts he had usurped. These formal gardens brought nature to the palace – but also brought nature to the city, and proved to be the inspiration for the public parks of the Enlightenment and the Industrial Revolution (Borsay, 2005).

This value of nature led to a flourishing of leisure activities designed to engage all people, rich or poor, urban or rural, with the countryside: angling and walking the banks of rivers, visiting shrines in woods and grottoes, and the first adventure tourism of mountain and hill climbing. Other humanists, such as Petrarch (himself a legendary climber of Mont Ventoux in France – though see Thorndike, 1943, for a critique of Petrarch's claim to have been the first modern climber), encouraged their readers to find Roman and Greek ruins from antiquity in the wider rural landscape, as well as in the towns and cities themselves (Mazzotta, 1993). The great freedom of belief and thought associated with Protestantism led to a number of large radical, religious movements emerging that drew on the romantic notion of the Divine in nature to nurture and encourage their Christianity in the woods (Lindberg, 2009). This was, supposedly, something more authentic to the early Christian church, its aboriginal practices and beliefs. Charismatic preachers took to the roads and spoke at rallies in fields; others emulated the early Christian hermits and lived frugal lives in the wilds (Ibid.). All this activity encouraged the growth of rural tourism and leisure, but rural leisure practices that were deemed more properly Christian: pilgrimage and discussion were replacing revelry and the sexual freedoms of feasts (Mazzotta, 1993).

As the rural landscape became more properly Christian, so the old pleasures of the woods were increasingly and literally demonized (Webster, 1982). The trend toward free thinking and freedom in leisure transformed medieval popular beliefs about witches and witchcraft. No proper historian has ever claimed that witches were actually making pacts with Satan. Charles Mackay, in *Extraordinary Popular Delusions and the Madness of Crowds* (2003[1841]), condemned witch-hunts as irrational, driven by superstition, passion and ignorance. Some historians have argued that there were some pagan traditions still extant at the time of the witch-hunts, but the evidence is weak (though see Ginzburg, 1992).

Keith Thomas, in *Religion and the Decline of Magic* (2003) suggests that popular superstitions about wise women and cunning men survived the Middle Ages. These men and women were effectively the healers in villages where there was no other authority. They worked inside popular Christian belief systems. Some of these also used folk magical ideas to curse people as well as cure them. It is these people (especially the women) who were mis-identified as witches.

Keith Thomas also suggests that the Reformation forced a change in attitude to the popular superstitions. On the Protestant side, the superstitions were proof of the work of the Devil in Catholicism. For Catholics in the Counter Reformation, the popular superstitions were dismissed as non-canonical, then heretical. Frances Yates (1964, 1972) has argued that the belief in witchcraft and the growth of the craze was connected to the rise of Hermetic knowledge and natural magic – occult philosophies that came into intellectual fashion after the Renaissance. These Hermetic ideas legitimized Manichean ideas about good and evil, and hence were connected by orthodox scholars to the medieval Devil-worshippers. These ideas also legitimized beliefs in magic and the power of magic.

Barstow (1988) has suggested that the witch-hunts were legitimized by the lower value of women in the late Middle Ages and early modern period. Women were at risk in a misogynistic, rural society, especially if they demonstrated any independence of mind or power. Barstow points out that most inquisitors and witch-hunters were men, and most of the people accused were women. Margaret Murray, in *Witchcraft in Modern Europe* (1921), estimated, by extrapolating from one local source, a figure of 9,000,000 executions of witches in this period. Barstow, making assumptions about the loss of records, says 100,000. The most accurate estimate, accepted by most historians, is that of 40,000 reached by Ronald Hutton (2006). Hutton suggests that neo-pagans and liberal historians have deliberately over-estimated the deaths, but he does recognize that the persecution of men and women as witches was linked to tensions between increasing freedoms in leisure and the constraints of local societal norms.

The Reformation and the retreat from work into rural spaces by some Protestant groups stemmed from communicative desires to find authenticity in nature. This desire was an echo of the later Romantic Movement of the nineteenth century, which, in turn, foreshadowed the rise of environmental tourism in the twentieth century. The rise of humanism was a reaction to the power of the Catholic Church. Competing notions of freedom, culture and morality challenged medieval Christian views of leisure. At the same time, however, rulers and states started to recognize the instrumental uses of leisure.

Guides to manners and Venice

As well as religious changes, Early Modernity is marked by the rise of bourgeois culture in the free cities and states of mainland Europe. The new elites from the free cities demanded lessons on how to act in a noble manner, so they could prove their distinction and civility (Bryson, 1998). At the same time, the old aristocracies were becoming more concerned with marking themselves out from the masses according to their tastes and morals (Tribby, 1992). These two trends led to the establishment of a refined, courtly high culture, with rules on gentility, decorum and distinction (Arditi, 1998).

Elias, in *The Civilizing Process* (1978, 1982) charts these trends and stresses the growth of inhibition and self-restraint, and the growth of privacy in the lives of individuals. Elias argues that these trends originated in courtly society of the late Middle Ages but grew with rise of bourgeois, urban middle classes, in early modern states such as Venice, France and England. Elias links this 'civilizing process' to the rise of absolute power in the monarchy. Successive kings moved to weaken feudal nobility, and legitimate use of violence became monopoly of the state. The outcomes of these 'civilizing processes' were a decline of violence for enjoyment, the banning of rough sports and leisure activities, and a decline in the public exercise of bodily functions (Arditi, 1998; Borsay, 2005). These outcomes can be seen in many examples of guidebooks on manners for would-be courtiers and members of the bourgeois high culture, where men (and, to a lesser extent, women) are told how to eat, how to drink, how to meet strangers, how to dance and what kind of music and sports are socially and culturally acceptable (Elias, 1978; Arditi, 1998). Members of the elite are also told how to play games and sports in a way that befits their status – Castiglione's *The Book of the Courtier* (1528, extract of primary source translated and cited in Orlin, 2009, p. 153) holds that the qualities of a Courtier should include a quasi-amateur ethic of playing for playing's sake, and a warning not to play the fool:

> To play for his pastime at dice and cards, not wholly for money's sake, nor fume and chafe in his loss; To be meanly seen in the play at chests [chess] and not over-cunning... Not to use sluttish and ruffian-like pranks with any man.

In the north of Italy, Venice stood as an independent trading market and a centre of bourgeois high culture. Like other cities in Italy, its freedoms had been challenged in the Middle Ages through the long period of political struggle between the Pope and the Emperor. Whatever its legal status, Venice was de facto an independent state throughout the late medieval and Early Modern period, ruled by an oligarchy of merchant families led by the Doge. With strong economic interests in the Mediterranean and the East, Venice acquired wealth and power to become, in the sixteenth-century and seventeenth-century, a cultural trend-setter for the rest of Christendom (Muir, 1986;

Burke, 1999; Horodowich, 2005). Venetian printing presses set the pace for the free exchange of ideas across Europe (Richardson, 2004), and Venetian artisans refined techniques for making industrial quantities of glass and guns (Norwich, 1982). Venice imposed cultural hegemony over much of northern Italy and the East, successfully challenging the Ottomans on the sea and, more prosaically, in the free market of luxury goods (Fenlon, 2007).

Although a Catholic city and province, Venice nurtured free-thinking individuals who were able to take advantage of the new invention of the printing press and the typeset book to promulgate ideas on a scale never seen before (Richardson, 2004). Alongside the scholarly translations of Classical texts, the style guides mentioned above, and the practical guides to husbandry and mechanics, were hundreds of other books that established reading as a serious leisure pursuit (Burke, 1999). Across Europe, rich individuals bought libraries of books, from the intellectual to the scurrilous, and reading became a leisure fashion for wealthy women who had little other private leisure pursuits. As well as being a leisure activity in itself, the reading of books allowed a common bourgeois public culture to emerge in this period – leisure practices and cultural trends learned from reading of them in books (Ibid.). So fashions for popular music, historical plays and for Italian clothing were transmitted by the medium of the printed word. The Protestant turn to individual study of the Bible in secular languages sanctioned the growth of literacy among the lower classes (Lindberg, 2009) – the initial religious demand to learn to read soon transformed itself in the urban lower middle-classes into a market for books as mindless entertainment: almanacs and horoscopes, biographies of famous jousting professionals (Crouch, 2006), tales of crime and scandal, and pornography (Eisenstein, 1983). These books popularized local peculiarities and tastes in leisure, a sign of the growing realization that leisure was something controlled, commercialized and instrumentalized. In some instances, classical fashions in leisure were re-invented. So Buckminster's *A New Almanac*, for example, devoted a section of its astrological advice to the best star signs under which the reader should use public baths (cited in Orlin, 2009, pp. 187–98).

Many Protestant cities overtook Venice as centres of printing, and although some Protestant rulers were as zealously censorial as the Catholic authorities, others allowed printers to print whatever books their reading public demanded (Eisenstein, 1983). The freedom of thought and comment made possible in Venice spread to the north, to the free provinces of the Netherlands and to England, where publishing created and sustained a desire for Italian bourgeois culture and leisure (Marfany, 1997; Burke, 1999). The printing revolution also saw the publication of dozens of travellers' stories, from merchants such as John Hawkins to ambassadors such as Sir William Roe (Orlin, 2009, pp. 235–41). These tales crated another literary leisure trend, the travelogue, which flourished with the growth of exploration of the New World, and the expansion of mercantile capitalism and Western European imperial conquest. The travelogue allowed the reader to visit far-off Mexico, India and China from the comfort of one's study (Benedict, 2002), a hot cup of coffee on the table a reminder of the growing trade imbalance that enriched Venice, England and the Netherlands. And related to the travelogue were the pilgrim and traveller guides to Rome, which allowed both faithful Catholics and humanist lovers of the Classical Age to follow in their mind the ways through the city's churches and ruins (Maczak, 1995).

In Venice, popular festivals associated with the holy days of the Medieval Church were transformed into expressions of bourgeois taste and distinction (Muir, 1986). Carnival, for example, had become in the late Middle Ages a way for the free and un-free poor of Christendom to subvert the ordinary systems of feudal control, through its licensing of sinful behaviour and anti-authoritarianism. Such behaviour was, of course, tightly controlled by the limited nature of Carnival – peasants were only ever kings for a day, after all (Scribner, 1978). However, in Venice, Carnival became something that the poor merely observed and cheered: from active participants, the poor became passive audiences for the elites who gathered to dance, to dress-up in their richest masquerades, and to eat and drink (Fenlon, 2007). Some elements of these festivals remained accessible to the masses (such as participation in boat races and processions – see Norwich, 1982), but the wealth of Venice trickled down only to the vendors and factory owners, the printers and teachers, the new bourgeois middle classes of the city (Fenlon, 2007).

This was a deliberate policy by the Doges: attendance at formal masquerades demonstrated that one had 'made it' into the ranks of the well-mannered and well-off elite; and the public display of ostentation told those who gathered to watch that they were in their proper station (Muir, 1986). The masses could be deceived into thinking they had the opportunity to act in such a way in their leisure life, if only good fortune would come their way, or if they would work hard enough for their masters. Gossip about the costumes worn by the elite must have been as depressingly prevalent as the celebrity pages of tabloids in today's media (Horodowich, 2005). For all the glamour and ostentation of the music, the dances and the costumes, in the Republic of Venice there were thousands of men and women who had few leisure opportunities, little leisure time, and little money to spend in that leisure time. Using the example of Venetian festivals and popular culture, then, we can see that leisure in the Early Modern period increasingly marked out a site where hegemonic powers used instrumentality to construct political cohesion.

Shakespeare's leisure lives

> FALSTAFF
> *By the Lord, thou sayest true, lad. And is not my hostess of the tavern a most sweet wench?*
>
> *PRINCE HENRY*
> *As the honey of Hybla, my old lad of the castle. And is not a buff jerkin a most sweet robe of durance?*
>
> *FALSTAFF*
> *How now, how now, mad wag! what, in thy quips and thy quiddities? what a plague have I to do with a buff jerkin?*
>
> *PRINCE HENRY*
> *Why, what a pox have I to do with my hostess of the tavern?*
>
> *FALSTAFF*
> *Well, thou hast called her to a reckoning many a time and oft.*
>
> *PRINCE HENRY*
> *Did I ever call for thee to pay thy part?*
>
> *FALSTAFF*
> *No; I'll give thee thy due, thou hast paid all there.*
>
> *PRINCE HENRY*
> *Yea, and elsewhere, so far as my coin would stretch; and where it would not, I have used my credit*
>
> (Henry IV, Part One, 1:2)

In *Henry IV, Part One,* the young Prince Henry – the future Henry V – is seen initially in the company of a rascal knight, Falstaff. This knight encourages the young prince to spend his leisure time in the taverns and brothels of London and to engage in acts of hooliganism and criminality. Later on, the young prince repudiates and plots against Falstaff in a scene (2:4) explicitly described as being staged in 'The Board's Head Tavern, Eastcheap'. Although this play is set a century earlier than the time it was written, the London it portrays, and the seediness of the city's night-time economy, must have been familiar to William Shakespeare. The famous Globe theatre, where Shakespeare's plays were performed, was in the red-light district of Southwark, alongside sites for bear-baiting and other violent sports (Linnane, 2007).

The end of the sixteenth century, and the beginning of the seventeenth, saw the establishment of a literary culture in England that is still considered the epitome of Englishness and a central component of the Western canon (Bate, 2009). Part of this literary culture – the plays of William Shakespeare and his contemporaries – is still lauded by critics and writers across the globe. Although something uniquely English, Shakespeare's plays speak of an essential human condition – of love, betrayal, conflict, tragedy – that is familiar to all of us. This literary culture, then, has become something analysed, taught, performed and critiqued anywhere that Western education has influenced. Shakespeare's plays are universal in their application, but grounded in the specific, literary, middlebrow culture of late Elizabethan and early Stewart England (Gurr, 2004). This culture had its origins in the Renaissance and the Reformation.

Political and religious changes in England in the sixteenth century saw the rise in the importance and influence of Parliament. There was an allied growth of the apparatus of the state and centralization of power (Ryrie, 2009). In 1521, Henry VIII attacked Luther and defended the Church in *The Defence of the Seven Sacraments* and was awarded the title *Fidei Defensor* by the Pope. However, in 1529 Henry, with Thomas Cromwell as his key advisor, started the break with Rome, instigated by the Church's refusal to allow Henry to divorce his first wife to be able to marry someone else who could offer him a son.

By 1534 the Act of Supremacy was passed. There were, of course, precursors to this break with Rome in England, such as Wycliffe and the Lollards (Lindberg, 2009); because of these earlier traditions, and because of the growth of Protestant preachers in England, there were enough priests and lords sympathetic to Protestantism to provide Henry with enthusiastic supporters. On Henry's death, Catholics struggled with Protestants for power over the throne. Under Mary, Protestantism was eclipsed, and Catholicism re-imposed, with predictably bloody results (Ryrie, 2009).

Under Elizabeth, there was an establishment (or reaffirmation) of Englishness, associated with rise of England as a sovereign nation resisting Catholic Europe. There was an attempt to reconcile the popular mood in favour of tradition and Rome with the 'fundamentalism' of the Puritans – despite a suspicion of Catholics and Catholicism, and some executions, there was no attempt to enforce Puritanism. Elizabeth established the Church of England, with its quasi-Catholic rituals, as part of the English state, with herself as supreme governor. She was excommunicated by the Pope (Pius V) in 1570. This led to an identification of Anglicanism with the Crown, and England; and Catholicism with Europe and the ruling crowns there. When Elizabeth died without an immediate heir, the English crown passed to the Stuarts of Scotland, and James I continued to defend Protestantism and the Anglican settlement against foreign interference and dogmatic Puritanism.

It is against this backdrop that Shakespeare – and his fellow playwrights Johnson and Marlow – wrote their plays. London was a huge, mercantile city, growing rich on England's importance in global trade markets and its connection to other Northern European states such as the Free Netherlands (Jardine, 2009). Actors, musicians, stagehands and playwrights belonged to professional companies, patronized and sponsored by the nobility, but dependent on the fickle masses for their audiences (Palmer, 2005). Theatres were built to give these companies permanent homes in the city, or on the edge of the city, where plays were seen as one form of attraction in a commodified, instrumentalized, leisure sphere. Inhabitants of the city could spend their money in taverns, brothels, ornamental gardens and menageries, watching blood sports, or attending a theatre to watch the latest play (Gurr, 2004). The plays themselves were of three basic types, two of which were reinventions of Classical Greek forms: populist histories, confirming Elizabethan English prejudices about the godliness of England and England's destiny; comedies, usually contemporary or loosely historicized farcical situations based on misunderstandings and resolved by some happy reversal of misfortune; and tragedies, histories that told some moral or allegorical tale of evil-doing and its consequences. In all these plays, just as in the small quote from *Henry IV*, there is a rich sense of Elizabethan and Stuart England's social and cultural life (Bate, 2009) – and that includes, of course, everyday and elite leisure patterns, practices and choices. So there were hunting, formal dancing and dinners for the rich and the nobility; taverns and rough sports for the free men; the domesticity of reading and making music for noble women; and the tightly controlled private domestic leisure of poor women, who only have enough leisure to talk to other women, and children.

In the comedies, in particular, late Elizabethan and early Stuart bourgeois leisure lives can be seen in their clearest detail. There are common themes in most of the comedies about romantic attraction and lust and love, of private desires overcoming expected duties. There is a reliance on double entendres, and a sense of knowingness about narrative perspectives and the agency of audiences (Gurr, 2004). There are courts, feasts and formal dances, young nobles wearing the latest fashions, and amusing songs (Burke, 1999). There are also more subtle themes – homosexual desire and homosociality in *The Merchant of Venice*, for example, or the taboo nature of playing above or below one's station in life mocked in *A Midsummer Night's Dream*.

The representations of Europe and the past in the comedies (and in *Romeo and Juliet*) are not anthropologically or historically correct, but they do show the contemporary fascination with northern Italy and Italian high and popular culture, and a strong acquaintance with the Classical Age. Shakespeare also gives us a clear sense of England's popular culture and leisure practices. His language is peppered with expressions from hunting and other field sports, and there are indirect references to other popular leisure activities such as story-telling. Indeed, much of Shakespeare's genius is the way he has combined the leisure habits of the bourgeois reader – who has read Hollinshed and other historians, as well as English translations of Roman and Greek texts (Bate, 2009) – with the folk narratives and symbols of medieval England's fireplaces and taverns. Shakespeare had absorbed much of England's oral culture, inherited the key, serious leisure role of story-teller, and transformed those long-told and elaborated stories into learned texts cribbed from the printed books of the Renaissance.

Shakespeare's plays have survived their time and context to become key texts of the Western and global canon. But even when they were being performed, there were those puritans who condemned the plays for their immorality, their lasciviousness and their un-Christian characters and situations (Kastan, 1986). The actors were all male, and the dressing-up of men as women was also something of which the puritans disapproved (Howard, 1988), though the proximity of the theatres to Southwark's brothels, and the bawdiness of the crowds, were more telling points in the puritan arguments. The English state at first encouraged the theatres and this flourishing of popular culture – and the themes of loyalty and Englishness in plays like *Henry V* served as effective instruments in the control of the city's masses (Gurr, 2004). However, as the English seventeenth-century progressed into the Stuart period, and bourgeois religious and social sensibilities increased, it was inevitable that the state would turn against the theatres and the playwrights. The value of the plays as propaganda, instrumental education and nation-building devices was nothing compared to their close connection with immorality and uncontrolled, dark and debauched leisure. When the Puritans took control of the English Parliament and the apparatus of the state, the leisure lives of everybody changed.

Puritanism and the English Civil War

Trends towards free thinking and liberty in English Early Modernity became more problematic in the religious and social struggles of the 1600s. This had an impact on people's leisure – both the view that leisure was something communicative and free, and the concern that leisure was something that had to be controlled by the state. We can see this by concentrating specifically on the example of the rise of Puritanism in England, its transferral to the New World, and the English Civil Wars.

Puritanism in England emerged out of the struggle over control of the English Church in the sixteenth century. The Puritans, those Protestants who believed the Anglican Settlement to be too close to Catholicism in theology and style, were a grassroots religious movement, with strengths in particular areas such as East Anglia (Spurr, 1998). They appealed to the new literate bourgeoisie and the rural poor and established a voluntaristic, active and communicative form of communal debate (Coffey and Lim, 2008). The plurality of religious belief across Puritanism led to tensions between freedom of expression and tolerance of dissenting opinions. This tension was present in opinions about traditional leisure activities, such as sports, music and dancing, as well as other leisure activities that were more obviously tainted with sin such as drinking, lechery and gaming (Spurr, 1998). For some Puritan leaders and writers, all leisure pursuits that were distractions from God were abhorrent; but others saw no harm in anything done in moderation and in the bounds of Christian morals (Bremer, 2009).

Puritanism spread to England's American colonies in the first half of the seventeenth century at a time when the Stuarts were backtracking from explicit Protestantism and wielding their autocratic power against the members of the English Parliament (many of whom were key Puritan leaders – see Spurr, 1998). Some of the more extremist Puritans declared that the fight to build communities of true Christians was lost in England – but the New England across the Atlantic Ocean provided fertile soil for such utopias. The Puritans established a number of colonies in New England, where they established autonomy from England and some measure of actual independence from the English church and state. These new colonies allowed charismatic preachers, ordinary men and women, and demagogues to establish moral codes based on local interpretations of what was acceptable in people's social, cultural, political and leisure lives (Kidd, 2005). As Bremer argues (2009, p. 57), these puritans:

> were not pleasure-hating, but they did try to place leisure and recreational activities within the framework of how they viewed the moral life. They believed that there was a proper place for leisure, which they saw as necessary to refresh and strengthen the individual, enabling him to return more effectively to the pursuit of his earthly and spiritual tasks.

So the American Puritans encouraged learning, exploration of the outdoors, but also sports such as nine-pin bowling, in their proper place (Kidd, 2005), but they did not tolerate blood sports, excessive alcohol use or sex outside of marriage. The Puritans controlled their colonies for much of the seventeenth century – but as the English Civil Wars came and went, and the Crown was restored in the home country, the control of the colonies was taken out of the hands of the Puritans by governors appointed by the English state.

The English Civil Wars were not the first civil wars in England, but they were the first fought with (early) modern weaponry, the first to bring devastation to towns and villages as well as soldiers in the field of battle. Charles I was not the first king of England to be killed by his enemies, but he was the first to be publicly executed following a formal decision. The cause or causes of the English Civil Wars have been argued over by many historians: Whigs, Marxists and assorted revisionists in particular. Whig Histories are liberal accounts of the progress of democracy, written in the nineteenth century but also in more recent times (Butterfield, 1968[1931]). They see the Civil Wars as the outcome of a long struggle between Parliament as defenders of freedom and absolutist Royalty. The Civil Wars are part of the wider Puritan Revolution paving way for religious toleration in the late seventeenth and eighteenth centuries. In these histories, Puritanism is the expression of traditional rights of the English people against arbitrary power. Twentieth-century historians such as Hill (1991) challenged the Whig historiography of this period. Hill drew on the work of Marx and the inevitability of class struggles from feudalism to socialism. For him, the Civil Wars were seen as a class war, between the aristocracy and the urban middle and working classes – a classic bourgeois revolution marking the and of feudalism. Puritanism becomes a bourgeois belief in individualism. More recent revisionist critiques of the Whig and Marxist accounts challenge the idea the Civil Wars were part of any long-term trend (progressive, class conflict) and stress the discontinuity between Tudor and Stuart England, the wider focus of the conflict, and the complexities of loyalties and individual lives (Hughes, 1998).

During the Civil Wars and afterwards, leisure became the site of an instrumental struggle over morality and meaning, between reformers seeing the work of the Devil in idle hands, and Monarchists who used the suppression of popular culture and leisure as means to gain support for their anti-Puritan campaigns. The Puritans who dominated the Government of the Protectorate enacted laws banning leisure activities associated with the Medieval Catholic world, such as friars and festivals on Saint's Holy Days (Hill, 1991). They also banned a host of sports and recreations on the holy day of the Sabbath, and banned sports and leisure activities such as bear-baiting, maypole-dressing and dancing, which offended their Christian sense of decorum. These bans were never enforced beyond some areas that were predominantly Puritan, and there is some debate among historians about the meaning and official nature of the bans (Bremer, 2009). But whether the bans were carried out, and whether they carried the force of execution beyond the rhetoric of Puritan generals, what is obvious is that the non-Puritan majority of England was coerced into self-censorship and restrictions in their everyday lives (Spurr, 1998). The great radical hope of the Civil Wars was expressed so eloquently in the Putney Debates of the Leveller movement, from which Lilburne's Agreement of the People (1647) set out the case for individual religious and moral freedom, and equality before the law (found at http://www.constitution.org/eng/conpur074.htm):

1. That in all laws made or to be made every person may be bound alike, and that no tenure, estate, charter, degree, birth, or place do confer any exemption from the ordinary course of legal proceedings whereunto others are subjected.
2. That as the laws ought to be equal, so they must be good, and not evidently destructive to the safety and well-being of the people.

Unfortunately, the instrumentally rational Puritan drive to suppress leisure practices, and limit choice and freedom in people's leisure lives, played into the hands of the Monarchist party, who portrayed the Stuarts as defenders of traditional English liberties and leisure (Hill, 1991). Public feeling turned to the Crown as a symbol of certainty and traditional values, against the wrathful modernity of Cromwell and the Puritans. Leisure was not the only area of life used in the struggle between the Puritans and the Crown, but it was one that all sections of English society could understand – whether it was right to get drunk, whether it was the role of the state to ban drinking, whether an autocrat like a Stuart king would be a defender of private liberties while controlling political freedoms.

The Restoration saw the return of theatres, brothels and drinking to the streets of London (Burke, 1999; Linnane, 2007). This period saw the emergence of the coffee house as a place for bourgeois conversation and fashion, and the emergence of the Royal Society as a place where gentlemen could discuss natural philosophy. In some ways, then, the Restoration created the climate for the Enlightenment – and nurtured some individual agency in the pursuit of leisure activities. But the price was a restriction of political liberty – the Crown and the state used these leisure pursuits to keep the public in check, and their hegemonic control secure.

Conclusion

Habermas (1989[1962]) describes how the jousts and other festivities associated with the nobility of Europe became increasingly associated with the palaces of the princes. This association defined an elite culture that was dislocated from the taverns, towns and tradesmen of the emerging European states. Through the early modern period of the sixteenth and seventeenth centuries, this elite sphere was subject to, and dictated by, the whims of the ruler. But in the establishment of autocracy in high culture, there was, in the same period, an inevitable reaction against such feudal submission. Those men (and the few women) who owed their wealth to their own capitalist endeavours, at first denied acceptance into the elite, soon found their wealth bought them status and recognition, as princes out-spent their land-based resources. Autocracy, then, gave way to a synthesis: what Habermas calls the good society. This 'good society' was both a part of the social world of the eighteenth-century Royal Courts, and a product of the rise (on the back of colonialism and industrialization) of the early modern capitalist economy. These, in turn, were connected to the rise of nation states and concepts of territory and power. With the freedoms and individuality associated with capitalism, trade and the emergence of power away from the Court, the good society flourished. And the good society allowed the creation of the modern public sphere.

> In comparison to the secular festivities of the Middle Ages and even of the Renaissance the baroque festival had already lost its public character in the literal sense. Joust, dance, and theater retreated from the public places into the enclosures of the park, from the streets into the rooms of the palace...Now for the first time private and public spheres became separate in a specifically modern sense...The authorities were contrasted with the subjects excluded from them; the former served, so it was said, the public welfare; while the latter pursued their private interests. (Habermas, 1989[1962], pp. 9–11)

This moment, the birth of the public sphere in the coffee houses of eighteenth-century London, Edinburgh and Paris, is also the high-point of European rationality: the Enlightenment. It is the time of radical political ideas such as secularism and republicanism; an age beginning with a conservative bookworm like Gibbon, and ending with the modern science, atheism and mathematical reasoning of Laplace. Until quite recently, men like Hume, Kant, Diderot, Voltaire, Rousseau and Lavoisier needed no introduction or footnote among the educated classes of the West: these were the *philosophes*, those who had rejected superstition and autocracy, and established reason and rationality as the only arbiters of truth. However, although these men – for the published and well-read writers of the Enlightenment were mainly men (Hankins, 1985) – wrote for a public that would discuss their work in public places, that public was not the same as the people: the public was the bourgeois classes, who saw themselves as intellectually and socially distinct from the popular masses.

9
Leisure in Modernity

At last, we arrive in the period on which leisure theorists and historians have traditionally concentred their efforts: the Enlightenment, the industrial age and the age of modernity (Borsay, 2005). In the first section of the chapter, I will use the well-known story of Benjamin Franklin using a kite in experiments on the effects of electrical discharges to introduce two related concepts: the leisured class of the Enlightenment, and the leisure activity of public science. I will discuss the key work by Habermas on the growth of the public sphere in the Enlightenment, exploring the rise of public discourse in periodicals in the United States, and the coffee houses of London. I will explore the idea of the Grand Tour as part of the education of leisured gentleman. I will argue there was a communicative relationship between leisure, reason and philosophy at this moment in European history, but the Enlightenment was also an age of dark leisure practices: as well as the coffee houses, Enlightenment London was home to other leisure pursuits such as drinking, gambling and prostitution. The rise of drinking as a mass cultural leisure pastime was linked to industrialization and globalization: the second section of this chapter explores the impact of both of these trends on leisure in the nineteenth century. I will explore the growth of tourism in tandem with the rise of European empires, the commodification and professionalization of leisure, the use of physical activity and physical culture in the establishment of nationalist ideologies, and the development of modern sports. I will argue that modern sport's origins are a curiosity, and that attempting to understand the history of modern sport without an exploration of similar modern, instrumental trends of commodification and professionalization in other leisure and culture forms is simplistic. In the third and final section of the chapter, an exploration of communicative resistance to modernity will highlight the development of communicative forms of leisure based on nostalgic revivals, folk movements and the Romanticism and Classicism of high culture. It will be argued that this resistance to modernity was, nonetheless, predicated on false notions of authenticity and tradition.

Kite-flying: Franklin, the enlightenment and gentlemen of leisure

One of America's most famous fathers of the eighteenth century, Benjamin Franklin, is remembered in folk mythology as a dabbler in public science. He is reputed to have flown a kite in a storm as part of his experiments and inquiries into the nature of lightning (Hankins, 1985). This story has passed into folklore and common knowledge about Franklin, and Americans today still know the story. As that well-known source of latter-day mythologies *Wikipedia* explains it, in a paragraph that, no doubt, is regularly cited in high schools up and down the United States:

> In 1750 he published a proposal for an experiment to prove that lightning is electricity by flying a kite in a storm which appeared capable of becoming a lightning storm. On May 10, 1752 Thomas-François Dalibard of France conducted Franklin's experiment using a 40-foot (12 m)-tall iron rod instead of a kite, and he extracted electrical sparks from a cloud. On June 15 Franklin may possibly have conducted his famous kite experiment in Philadelphia, successfully extracting sparks from a cloud, although there are theories that suggest he never performed the experiment. (http://en.wikipedia.org/wiki/Benjamin_franklin, accessed 23 August 2010)

This story encapsulates many things of interest to a scholar of leisure. First of all, there is the kite, evidence that such things were so common-place among late eighteenth-century Americans, and their nineteenth-century myth-making equivalents, that no other explanation is necessary. All American children – or rather, most likely boys – must have enjoyed building and flying kites, and to see a renowned politician and philosopher doing the same, one dark and stormy day, was not so strange as to bring ridicule on him. Secondly, this story demonstrates the existence of the gentlemen of leisure in the eighteenth century: that class of wealthy men who had free time from income-generation to be able to pursue their intellectual and leisure interests (e.g., sports such as hunting, or reading books). Thirdly, it shows that such gentlemen of leisure were educated, and driven by their education (whether formal study, or self-learning) to pursue scientific and philosophical inquiry as a leisure activity. Fourthly, the story encapsulates the period of the Enlightenment: the sharing of ideas in a public sphere, the testing of theories, and the involvement of the new leisured class in the construction of knowledge (Habermas, 1989[1962]; Borsay, 2005). Franklin's ideas would have passed from America to Europe through the exchange of letters, the publication of papers, and informal discussions in coffee houses and other public spaces, where literate men and women could meet in their leisure time. Not all exchanges in such public spaces were as worthy as the analysis of Franklin's ideas about electricity: public life in cities such as Paris and London was, in the eighteenth and early nineteenth centuries, soaked in alcohol binges, fights and debauchery (Melton, 2001). Nonetheless, the Enlightenment was a new turn in the history of ideas.

Habermas returns to the Enlightenment origins of the bourgeois public sphere in the second half of *The Structural Transformation of the Public Sphere* (1989[1962]), where he draws on the writings of a number of nineteenth-century political and social theorists to account for the transformation of the public sphere of the Enlightenment. In discussing the post-Enlightenment reactionary position of the bourgeois public against the masses, framed by the terror of the French Revolution (when the *sanculottes*, the poorest classes, led the mobs and, for a moment, held power over the bourgeoisie) and its Napoleonic coda, Habermas cites freely from the work of nineteenth-century German political philosopher Wieland (1857, 32: 191–218, cited in Habermas, 1989[1962], p. 102 and n. 50). Public opinion, according to Habermas (Ibid., p. 102) originated from people who had education, knowledge and understanding, reading periodicals in the United States or debating in coffee-houses in European cities. So, referencing Wieland, Habermas argues that this public opinion 'spread "chiefly among those classes that, if they are active in large number, are the ones that matter". Of course, the lowest classes of the people, the *sanculottes*, did not belong to them, because, under the pressure of need and drudgery, they had neither the leisure nor the opportunity "to be concerned with things that do not have an immediate bearing on their physical needs."'

For example, one of the developments in formal tourism in the eighteenth century was the establishment of the Grand Tour. Young men from elite families in Europe and America were sent on long vacations to explore the sites of old Europe: of Italy, and latterly, of Greece (Black,

1985; Chaney, 2000; Paul, 2008). These tours followed itineraries established in published guidebooks. An industry of guides, hostels and carriage companies was created to cater for the increasing normality of the Grand Tour, with companies competing with each other offering more and more lavish accommodation and travel. Sometimes, the tours included spells of formal education, but more often the tourists educated themselves by visiting the ruins of the classical world, or the holy places of the Roman Church (Paul, 2008). Such worthy heritage tourism was the purpose of the tours, no doubt, in the mind of the families that paid for their sons to travel; in the mind of the sons, the remains of the Forum may have taken second-place to the prostitutes in Rome's brothels. Whatever the motivation and the object of travel at any given moment, what the fashion for the Grand Tour shows is that the creation of the bourgeois public sphere allowed those who could afford it an enormous increase in leisure time – and leisure activities of whatever moral stamp.

Here we can see that leisure time, the ability to control one's work and resources to support time where one is free to think, read and discuss, is crucial for communicative rationality and action. Habermas (1989[1962]) expands on this point when paraphrasing Marx's normative programme for social and economic change. Marx, identifying the inequality in the bourgeois public sphere of power, autonomy and hence freedom, came up with a model, discussed by Habermas, for a public sphere in which such power is more evenly distributed. This, of course, is a utopian dream, but Habermas uses it to reflect back on the actual development in the nineteenth century of the bourgeois public sphere. Here, again, leisure is a central part of the analysis.

> The autonomous public, through the planned shaping of a state that became absorbed into society, secured for itself (as composed of private persons) a sphere of personal freedom, leisure and freedom of movement. In this sphere, the informal and personal interaction of human beings with one another would have been emancipated for the first time from the constraints of social labor (ever a 'realm of necessity') and become really 'private'. (Ibid., pp. 128–29)

So there was a communicative relationship among leisure, reason and philosophy at this moment in European history, but the Enlightenment was also an age of dark leisure practices: as well as the coffee houses, Enlightenment cities like London were home to other leisure pursuits such as drinking, gambling and prostitution. The development in the late eighteenth century of industrial factories producing beer and gin led to a proliferation of drinking houses in London. The factories could operate on a commercial scale because of the increasing application of technological advances, such as the use of steam engines, to production (Hankins, 1985). Gin, in particular, proved cheap to make on an industrial scale once the technology of distillation allowed it, and, as a consequence, it became the subject of the first modern moral panic (Warner, 2003). Because of the cheap retail price and the high alcohol content, gin became associated with poor leisure choices: drunken idleness, cockfighting and bear-baiting, casual sex, gambling, fighting and petty crime. Drinking, in London, had become a mass cultural pastime (Warner, 2003; Borsay, 2005; Linnane, 2007), and although the state tried to legislate against it, alcohol became a fixed part of the urban landscape, not just in London but in most other industrial cities in Europe.

Progress or disenchantment? leisure and the nineteenth century

The rise of drinking as a mass cultural leisure pastime was linked to industrialization and globalization: the second section of this chapter explores the impact of both of these trends on leisure in the nineteenth century. I will explore the growth of tourism in tandem with the rise of European empires, the commodification and professionalization of leisure, the use of physical activity and physical culture in the establishment of nationalist ideologies, and the development of modern sports.

Weber (1992[1922], 2001[1930]) famously describes the loss of magic and enchantment in the world brought on by modernity. Increases in big business and rationalization were leading to individuals feeling a loss of control. But Weber's pessimism was not matched by others. I want to challenge this myth about a simplistic link between modernity, progress, rationalization and the concept of disenchantment by looking at the work of the American radical economist Henry George. Henry George was a supreme optimist and author of *Progress and Poverty* (1979[1879]). He believed Social Darwinist ideas to be mistaken: we could all be middle class and wealthy, free from the trap of Malthusian despair. His was a liberal view of capitalism and progress, but combined with a moral concern with laziness.

George reflected the views of most people in the late nineteenth and early twentieth centuries – the beginning of modernity, the early years of modern sport and leisure. Hofstadter identified him as a vocal opponent of those (such as Herbert Spencer) who linked social Darwinism to capitalism (Hofstadter, 1955), and both Ruse (1996: p. 199) and Fichman (1997: p. 109) have shown that *Progress and Poverty* was an influence on Alfred Russell Wallace's later thinking about progress. Fichman also cites a letter from Wallace to Darwin, urging him to read *Progress and Poverty*, and a reply from Darwin that he had ordered a copy, but that political economy had a disastrous effect on his mind when he read it (Fichman, 1997, p. 109, citing the correspondence of Wallace collected and published by Marchant (1975, pp. 260–261). But George's actual theories have been largely ignored or overlooked by historians of leisure and popular culture, and left to be argued over by political and economic historians who concentrate on George's political career in America, where he was a champion of the labour organizations and the inspiration for radical movements for land and market reform (Post, 1930; Barker, 1955).

It is useful to examine *Progress and Poverty* to see what notions of progress George had, and how they related to his notions of evolution, science and economy. *Progress and Poverty* was an instant bestseller, and its influence on debates within the Liberal Party in England and the liberal movement in America has long been established (Lissner and Lissner, 1992). Books and papers about Henry George are usually hagiographic (Post, 1930), driven by contemporary political reform movements such as the Henry George Foundation or attempts to contextualize or criticize his economic arguments (Cord, 1985; Hellman, 1987). I do not intend to enter into a historiographical debate, nor do I wish to produce a textual history of how George came to write *Progress and Poverty*, but some sort of context is needed.

Henry George was not university educated, but he did get training as a printer, which allowed him to become printer for the *San Francisco Post*, then a reporter, then a prosperous and successful newspaper proprietor in his own right (Post, 1930; Barker, 1955). George was, in this sense, a member of the growing bourgeoisie in California, with connections to the political class and leading businessmen who had invested in his newspaper. *Progress and Poverty* was George's attempt to reconcile the industrial growth of the American nation with the rise of social problems associated with poverty, and to offer a solution to ameliorate the situation. He began by invoking the names of two scientists from the Enlightenment – Franklin and Priestley – and asking whether either of them could have seen, 'in a vision of the future, the steamship taking the place of the sailing vessel... could he have heard the throb of the engines that in obedience to human will, and a satisfaction of human desire, exert a power greater than that of all men and all the beasts of burden of the earth combined?' (George, 1979[1879], p. 1).

In recognizing that the world had changed and improved in terms of human ability to control it, George was identifying this change with science, and identifying himself with Franklin and Priestley – both of them provincials, and both, in their way, radical scientists. This industrial growth – the creation of the railways, the standardization of time, and the establishment of Western empires and capitalist hegemonies, led to the development of modern tourism (Borsay, 2005). Wolmar (2010) has pointed out the rise of elite tourism in the spread of rails and steam-packet companies across the world; Walton (2000) has shown how trains allowed a huge rise in leisure time and opportunities for the working classes of England. Both rich and poor in the developed West could take advantage of increasing wealth, freedom and technologies, all based on the appropriation of riches from colonies, and the abuse of fellow humans in the slave trade. Industrial growth gave the West the technological power to shrink the size of the globe, to bypass the Ottoman Empire in pursuit of Indian and East Asian markets and to enforce power through military force (Hobsbawm, 1988, 1989). Yet, despite the industrial growth, according to George there was poverty and distress everywhere, and 'beneath all such things as these, we must infer a common cause ... that we call material progress' (George, 1979[1879], p. 3).

For George, the solution to the problem of progress and poverty could only be found through the principles and methods of Political Economy, 'the science that seeks to identify, in the sequence of certain phenomena ... cause and effect, just as the physical sciences' (Ibid., p. 6). *Progress and Poverty*, then, is intended to provide a scientific answer and is a contribution to the growing scientific discourse of the late nineteenth century. George started from basic principles and claimed to be following a rational method of assessing the evidence, 'taking nothing for granted' (Ibid., p. 7). George's understanding of political economy was based around a reinterpretation of labour, capital and wealth, which lead him to discuss whether there is enough land, enough resource, for continual economic and social growth. He rejected the application of Malthus' famous argument to human society (Ibid., pp. 40–60). George argued that 'Man' is different to animals because he is able to provide his own subsistence through additional work, and 'give play to the reproductive forces' (Ibid., p. 53). For George, poverty was not the fault of over-population, because more people meant more labour and more wealth, which could be gained from the land more efficiently by the industry of more hands. Rather, poverty was the fault of the economy, which did not recognize the value of land and the distinction between use and ownership (Ibid., p. 159). It was a failure by politicians to recognize the weakness in *laissez-faire* capitalism and its assumption about continual progress and the natural competition between races. It was to the latter assumption that George addressed his chapter on the law of human progress, which dismisses the law of progress understood through evolutionary theory:

> The prevailing belief is that the progress of civilisation is a development or evolution, in the course of which man's powers are increased and his qualities improved by the operation of causes similar to those that are relied upon as explaining the genesis of species, namely the survival of the fittest and the hereditary transmission of acquired qualities. (Ibid., p. 186)

George was not arguing against evolution – the survival of the fittest – as a tool to explaining 'the genesis of species'. But he was against a reading of human history that suggested human progress was through a linear process of evolution. In this respects, his arguments echoed the opponents of progressive history discussed by Bowler who were ready to accept a cyclical version of history, inspired by Romantic historiography (Bowler, 1989, pp. 8–10; Cunningham and Jardine, 1990). George continued his argument on the uniqueness of Man's progress by asking his readers to look at the rise and fall of ancient civilizations, and at the causes of their falls: economy, that is, an economic system that fails to separate use and ownership of land, not over-population. As he concluded:

> Nothing could be further from explaining the forces of universal history than the theory that civilisation is the result of a course of natural selection which operates to improve and elevate the powers of man. (George, 1979[1879], p. 187)

To understand George's moral motivation for making the claim, against the Social Darwinists, that Man was not subject to Malthusian growth and the 'course of natural selection', it is necessary to look at his concluding chapter. George was keen to position himself as a scientist, an authority and a member of the imagined community of scientists. For him, his political economy was a science that provides laws, 'the laws that govern the production and distribution of wealth' (Ibid., p. 217), which show that 'the want and injustice of the present social state are not necessary' (Ibid., p. 217). George offered Man a way to true progress because, as he claimed:

> Economic law and moral law are essentially one ... the great fact that science in all its branches shows is the universality of law ... whenever

> (the scientist) can trace it, whether in the fall of an apple or in the revolution of binary suns... the law holds good. (Ibid., pp. 217–18)

George was aligning his political economy with the physical sciences of physics and astronomy, but it was not just the certainty of these sciences from which he wished to benefit. Ultimately, George recognized that physics revealed something of the mystery of the universe, the mystery behind the regularity of laws. Just as physics shows us something about the great mystery of the life beyond this one, he argued, so his law of human progress told us that this progression could only be understood as the progression of the individual soul (Ibid., p. 218–19). This land is bountiful, as George has shown, and there is an economy that will make Man's society just. But, for George, life must have a purpose, a progress: life 'seems only intelligible as the avenue and vestibule to another life' (Ibid., p. 219).

In some respects, *Progress and Poverty* is an infuriating text. Its economic arguments are dense, and the style recalls Whewell's method of consiliences, the inductive interlacing of evidence and assertion that Ruse has argued Darwin adopted (Ruse, 1979). But by looking at George's aims, and arguments about progress, we can see that he is against the application of the theory of evolution to Man because of Man's unique talents, purpose and destiny. In all other respects, George is anxious to legitimize his political economy as a science, and if his purpose is ultimately theological, the immediate concerns about poverty are real enough for him to ensure that his argument is, as he understands it, scientific. Hence, his invocation of Franklin and Priestley, his seeming acceptance (or, at least, his silence in not attacking it) of the theory of natural selection, of his unsubtle equating of economic and scientific laws: like Marx, George considers the question of economy to be the most important scientific question of the age. Unlike Marx, George's optimistic belief about progress and capitalism fitted the mood of the West in the age of Empire and Industry: reading his book, one could see a happy future of worthy leisure. Science and technology were changing the world and our view of it (Lightman, 1997): evolution and geological time; steam engines, transport and industry; the telegraph, telephone and radio; the laws of thermodynamics; and the industrial exploitation of chemical processes and electromagnetism. The success of science seemed to be linked to the success of capitalism. Political power was globalized, and free trade became the goal of modern capitalism. And with capitalism and science came the political beliefs of liberalism: protection of the individual; free speech; privacy and public space; and free leisure time (Borsay, 2005).

Across the globe, but especially in the West, these changes in society linked with modernity resulted in the commodification and professionalization of leisure: railways, for example, not only allowed the two coasts of the United States to be reachable from each other, but the railways built through the Rockies spurred the growth of outdoor tourism, the building of spas and the spread of ski resorts (Wolmar, 2010). In Europe, the use of physical activity and physical culture in the establishment of nationalist ideologies (Hobsbawm, 1992), and the development of modern sports, were outcomes of industrialization. Again, professionalization was a consequence of industrialization and technological change. The rise of the urban working class and the successful negotiation of Saturday afternoons free from work (Holt, 1989) allowed capitalists to develop leisure industries based on outdoor pursuits (Snape, 2004), drinking and gambling (Greenaway, 2003), and sports spectating (Collins, 1999). Wolmar (2010) claims the creation of professional team sports in Europe, with large crowds of passionate fans, could only be a consequence of the shrinkage of journey times and the standardization of time associated with the spread of the railways.

Modern sport

In 1908, reflecting on the increasing importance of the Modern Olympic Games in the wider geo-political landscape, their founder Baron Pierre de Coubertin expressed his hope that the Games were having 'an influence, which shall make the means of bringing to perfection the strong and hopeful youth of our white race' (De Coubertin, cited in Carrington, 2004, p. 81). For De Coubertin, the Modern Olympics were beyond nationalism, but not beyond the symbolic boundaries of early twentieth-century notions of race. The Games, as King (2007) has argued, were designed and promoted by an elite section of white European society at the end of the nineteenth century as a means of preserving and promoting elitist ideas of belonging and exclusion. The rhetoric of the open playing field symbolized by the Olympic rings masked the reality of sport's role as the maker (and marker) of racial difference (Mangan, 1981, 1985, 1995; Mangan and Ritchie, 2005): sport made white men fit to serve the engines of commerce and empires. One hundred years on from De Coubertin's reflections, sport is still part of the geo-political landscape, but the Summer Olympics in 2008 – hosted as they are by China – and the multi-ethnic character of the upper echelons of many professional sports might be suggestive of the irrelevance of modern, Western sport's racist origins. But racism, conceptualizing and constructing difference in terms of racial categories, is still part of the language and logic of sport: in the way people write about sport (Carrington and McDonald, 2001; Hylton, 2005; Burdsey, 2006); in the way decisions are made about the funding of sport (Spracklen, Hylton and Long, 2006); and in the way cultural differences in sport become biological differences.

As the century drew to a close, the Other became associated more with ideas of impurity and degeneracy. Concerns about the effects of unchecked capitalism and the unchecked Mathusian growth of humanity (Gould, 1997) combined with biological arguments around design and evolution. Although Darwin was cautious about applying the theory of evolution to modern society, this caution was dropped by his successors. Galton and others read *The Origin of the Species* as a moral tale about the declining birth rates of the middle classes and the growth of the working class. Social Darwinism became a movement that promoted the survival of decent middle-class society through eugenics: the breeding of good middle-class 'blood' or, alternatively, the restriction of breeding of the working class, foreigners and other undesirables (Kohn, 1995).

At the same time as scientific racism was being used to explain and justify inequalities at home and in the empire, the English middle classes were undergoing a cultural revival associated with healthy living and moral rectitude. Movements such as Arnold's 'Muscular Christianity' made an explicit link between the moral fibre of the ruling classes and physical activity, and sport became a way of making men fit to run the empire and run the capitalist system (Mangan and Ritchie, 2005). The modern sports of the modern Olympics were developed and codified by the ruling classes of Western Europe, and the exclusionary nature of amateurism – enshrined until the end of the twentieth century in the Olympic Movement and many middle-class sports such as rowing, cricket and rugby union – was testament to the hegemony of the ruling classes (Dunning and Sheard, 1979). However, sport was seen also as a way of improving and moralizing the working classes. The extension of the muscular Christianity movement into working-class communities was an attempt to combat degeneracy and immorality. Collins and Vamplew (2002) have shown the historical relationships among sport, alcohol and working-class licentiousness, and middle-class attempts to control and moralize sport. Many sports clubs and sports associations in the UK were first established by church groups or men of the church (Dunning and Sheard, 1979). So sport was a part of the social Darwinist enterprise, a means of improving nation, race and class – an expedient way of measuring the fittest and second-guessing the evolutionary winners but also a way of responding to the realization that nation and race were not evolutionary fit 'enough' to be those survivors. Almost by accident, the middle-class obsession with morality and the survival of the white race led to the creation of professional sports: where winning was everything, where the best of the nation were selected to represent the nation against other nations, and where the involvement of capitalist processes and professional interests led to the globalized, commercialized, sports business we recognize today (Horne, 2006).

Rugby league is a version of rugby football traditionally played mainly in the north of England, eastern Australia, southern France and New Zealand (though it is increasingly played in more countries – see Spracklen and Fletcher, 2010). The game split from rugby union in 1895 when a number of northern English clubs fell out with the Rugby Football Union over payments to players. Historically, rugby league was viewed as a 'professional' version of rugby: although the top leagues in England and Australia were and are professional, most rugby league clubs and leagues were and are amateur (and top rugby union clubs have been professional since 1995). The game is played thirteen-a-side and has more open play and fewer technicalities than rugby union. In Spracklen's *Playing the Ball* (1996), I conducted ethnographic research on rugby league (and union) in a northern English city. I found that traditional rugby league localities (communities?) were in decline or gone altogether. They had been replaced by an imaginary community, partly symbolic, partly imagined and invented as rooted in some (rose-tinted) 'reality' of the past: the myth of the Split; gritstone; gritty men; and Northernness – white and working-class. Of course, the communities in the past were also imaginary and imagined, in the sense that the Split gave them a myth and an invented tradition of working-class resistance, whose whiteness and maleness were refractions of the imaginary (re) invented throughout the age of urbanization and industrialization. This refraction still exists in rugby league; for example, the big Cumbrian forwards, big because they were bred in the steel works of Workington and Whitehaven, hardened in the rain lashing that coast ... exotic others to the normal northerners of Yorkshire or Lancashire. The men of the north (imagined in rugby league) come from particular elements of the working class: small, one-industry towns; or particular one-industry districts in the bigger cities. I have argued elsewhere (Spracklen et al., 2010) that postmodernity's fragmentation imprints on this past a unity and cohesiveness that it lacked.

The work of Tony Collins (1999, 2006) reflects the new orthodoxy of opinion internal to rugby league. Collins views Rugby league as a genuine working-class social movement (movement of resistance), situated in the leisure lives of its white, northern, working-class participants. In the work of Collins, as in the work of non-academic sports historians, rugby league becomes the north – the game's identity in the UK is fixed by northern England: its landscape, industry, housing and communities (such as the use of the Northern Union as a psychogeographical entity in twentieth-century English popular culture). Having previously been the writer of the alternative historiographical challenge to an official version, Collins' historiography of black involvement (in Melling and Collins, 2004) has become the new orthodoxy.

As my previous research into rugby league progressed, the idea that the people I spoke to used 'the game' to identify with this ghost of the past, this *idea* of what it meant to associate with a rugby league playing district, became very important, as values in 'the game' were conflated with the values of the working-class localities as *they remembered them* (Spracklen, 1996; Spracklen and Spracklen, 2008; Spracklen et al., 2010). This process resulted in what Anderson (1983) calls imagined communities, where historical invention has resulted in a cohesive structure for legitimizing a sense of community in the present. This was a process of reinvention of the past (Jenkins, 1991), inventing traditions that justified the values of the present (Hobsbawm and Ranger, 1983). As Eddie Waring put it:

> It's as North as hotpot and Yorkshire pudding. It's as tough as teak. It's rugby league – a man's game if ever there was one. Someone once said of rugby union, 'A game for ruffians played by gentlemen' ... Gentlemen have played rugby league. Gentlemen still do. But the hard core of rugby league players, with their cauliflower ears, their broken noses, their busted and bruised bones, would rather be called, to use a three-letter word, MEN ... It's a down-to-earth game played by down-to-earth people. Good people. Solid people. To use that three-letter word again, MEN. (Eddie Waring, 1969, flyleaf: emphasis in original)

For many years Eddie Waring was the voice of rugby league in the United Kingdom, an icon of the nation, whose particular West Riding pronunciation of the phrase 'and it's an up'n'under' when describing a high attacking kick (now called, in more masculine terms, 'the bomb') was mimicked by comedians everywhere. Eddie, the BBC's rugby league commentator, became a well-known and much-loved figure, with appearances on the *Morecambe and Wise Show* and a job hosting *It's A Knockout*. However, his fame seemed to eclipse the game of rugby league which he supposedly commentated on, and in the eyes of many rugby league fans Eddie was a northern caricature straight out of the Music Hall who did the image of rugby league no favours. His image of 'the game' was inextricably linked with notions of masculinity, of community, of class and of 'northernness'.

Modern sport's origins are a curiosity. But it can be seen that it was a product of the late nineteenth-century tension between different urban classes, and a product of wider pressures on leisure and Western society. It could not have appeared at any other time, or in other place in the world: it needed technological change, an urban working class given some freedom and wealth, and the freedom of industrial capitalism. Attempting to understand the history of modern sport without an exploration of similar modern, instrumental trends of commodification and professionalization in other leisure and culture forms is simplistic: too many histories of sport become Kiplingesque 'just-so' stories taken out of the context of instrumental pressures on Western, nineteenth-century leisure at the height of modernity. Modern sport was just another leisure site – like modern tourism, like the alcohol industry – where Habermasian instrumentality met a decreasing, but still significant, amount of communicative action.

Resistance through rituals: the invention of authenticity in folk tradition

In the advance of industrialization in the nineteenth century, and the globalization of Western modernity, leisure became a place of resistance for subaltern and other marginalized communities, the poorer classes and women. In the United States, slaves and former slaves developed informal leisure activities within the limitations of their servitude: musical structures and styles from the Church and white Europeans merged with half-remembered or reconstructed African rhythms to create the foundations for twentieth-century gospel, jazz and blues (Bennett, 2001). In Japan, caught between a pursuit of the modern embraced by its urban elite and the defence of native traditions, martial arts became codified as sports, acceptable to both Westerner-looking classes and the old generation concerned with feudal and religious correctness (Saeki, 1994). In India, cricket, the game of the British Empire, was co-opted by subaltern communities as way of demonstrating their ability to be both British and Indian, or as a way of claiming the imperial sport for their own identity-making purposes (Bose and Jalal, 2003). In France, women fought for the right to access beaches and other bathing areas (Hobsbawm, 1989). In Egypt, Islamism developed as a modern ideology against Westernization partly due to the increasing democratization of the city streets, the choice of debate as a leisure pastime, and the growth of public spaces not controlled by Ottoman laws and guilds (Hourani, 2005).

In the United States and across many other countries where free speech was not restricted, the growth of periodicals allowed writers outside of the political elite to express opinions (Briggs and Burke, 2009). These magazines and newspapers created niche audiences, readers who felt part of a wider community of sports fans, pigeon-fanciers, socialists or amateur geologists. More generally, regional and national newspapers established a sense of belonging to a public sphere of news and opinion: they created a sense of popular culture, letting their readers know which books to read, or plays to attend (Borsay, 2005; Briggs and Burke, 2009). The media in the nineteenth century allowed nationalist sentiments to coalesce around populist senses of history, myth and use of languages, allowing, for instance, nation-states to emerge in South and Central America around shared notions of cultural and historical belonging (Hobsbawm, 1992); or white Australians to develop an ambivalent relationship with Great Britain through the establishment of regular sporting tests (Williams, 1999).

Public and popular sentiments of national pride led to the argument in the public sphere that modernity, industrialization and globalization were leading to the loss of authentic, native, national cultures (Anderson, 1983; Hobsbawm and Ranger, 1983; Hobsbawm, 1992; Baycroft, 1998; Smith, 1998; Gellner, 2006). This was articulated as a loss of authentic leisure pursuits that were supposedly bound to the rural landscape, the deep history of the place, and the 'blood' of the nation's folk. Academics and popular writers argued that modern life had become ersatz, inauthentic, and only by returning to its roots could a culture find meaning and purpose (Baycroft, 1998). This, of course, was both a conservative ideological turn and false one, for authenticity was an impossible goal, in leisure or in any other aspect of life (Anderson, 1983). Nonetheless, many people tried to find some authentic, folk tradition which they could resurrect, or something they could (re)create. In the Ottoman Empire, slowly decaying in prestige and power, there came a flourishing of traditional, classical Ottoman music and poetry (Hanioglu, 2008), as well as more narrowly Turkish resurrection of wrestling (Stokes, 1996). In Russia, the ruling elite turned from an admiration of French culture to a Slavophilia that promoted a narrow interpretation of authentic Russian pastimes, dances and music – reducing Jewish influences on those 'traditions' as well (Lee, 2005). In the newly created Germany, hiking in the Bavarian mountains, for example, became a way of connecting with an imagined, rural past for the urban middle-classes of the Prussian cities, even though the hinterland of northern Germany had no mountains of note (Baycroft, 1998). German concerns with the Aryan correctness of their *volkisch* roots, of course, soon led to ideologies of race, masculinity and nation being intertwined with military training, gymnastic sports and, by the twentieth century, the healthy living and folk music of the Hitler Youth (Kater, 2006).

In England, folk revivals became popular following the publication of Frazer's *The Golden Bough: A Study in Magic and Religion* (2004[1890]). This monumental work of anthropology and speculation argued for the survival in rural parts of Europe of traditions of nature worship into pre-modern times, witnessed in the cycle of festivals, folkloric practices and leisure pursuits that were being extinguished by the rise of cities and industry. The book explained to its English readers why they felt nostalgia and loss for a golden age of village-life, tied to the annual cycle of seasons, the long nights of winter, the death and rebirth of the Sun. Frazer suggested that, once upon a time, social cohesion was maintained through participation in rituals that ensured the successful passing of seasons. These rituals were part of our leisure lives, fixed events in the year, where our full participation was essential if the rituals were to succeed. Frazer did not believe in magic, but he did believe that his idealized peasants were bounded by such superstitions, and the rich diversity of fairs and feasts were more than just an opportunity to drink, eat and subvert convention. His richly-evidenced book persuaded many people in England and Europe to seek out the remnants of folk leisure in far-off villages. Some people decided to record songs sung by travellers and others living in rural places, with suitable removals of modern influences from music hall (Francmanis, 2002). Books were published about dances and sports that had survived industrialization, such as the so-called folk football games of Ashbourne or Workington, or the hobby-horses of Padstow, which were said to be pagan in origin (Hutton, 2006). Others read local histories to try to find traditions that had died out, which could be revived. This led to the (re)construction of English folk dance in the form of Morris and mummery, a process that continued through the twentieth century (Garry, 1983). These folk revivals spread to other parts of the world – becoming ways of appropriating aboriginal cultures that had been wiped out or marginalized by colonization, as in the Americas; or ways of defining nationalist belonging in areas of conflict and challenge, such as the Basque country or Hungary (Gellner, 2006).

What these forms of resistance and revival share in common is a quest for an authentic, communicative leisure experience, which empowered or legitimized particular social groups. The quest has been clearly articulated by MacCannell (1973, 1976) in his work on modern tourism and pilgrimage:

> The touristic way of getting in with the natives is to enter into a quest for authentic experiences, perceptions, and insights ... Tourists make brave sorties out of their hotels hoping, perhaps, for an authentic experiences, but their paths can be traced in advance over ... what is for them increasingly apparent authenticity proferred by tourist settings. (MacCannell, 1973, p. 602)

The morality of authenticity elides smoothly into a Western, middle-class sensibility of culture: the authentic is good because it runs counter to the homogenizing tendencies of globalization, because it encourages diversity and respect and cultural heterogeneity. MacCannell sees authentic cultures as existing, in a Gofmannesque sense, backstage. But, of course, for MacCannell, authenticity is not something that can be grasped by tourists in any objectively *real* sense: the authentic is something 'which is for them', the tourists, merely *apparent*.

The search for authenticity and belonging in the late nineteenth century led to mythology, purity, then, eventually, extremism and violence. But the genesis of these forms was situated very clearly in the late nineteenth-century shift to nation-state building, which Hobsbawm (1992) clearly identifies, whether nationalism in the name of a new nation-state or a proto-nationalism associated with a region or a particular subaltern community, or a folkish notion of national culture. These forms of resistance and revival were reactions against, or appropriations of, modernity and modern forms of leisure: sport, tourism, popular culture. What modernity is, and what people thought it was, is the subject of the next chapter.

10
Leisure in Historiography

In this chapter, I will explore the theme of leisure in writers from the Early Modern period onwards. The first half of this chapter will juxtapose primary sources such as the work of Milton, Gibbon and Mill with modern historiography of leisure (see, for example, Borsay, 2005), both to examine their ideas about the purpose of leisure and to provide a sustained engagement, following the Habermasian critical lens, with contemporary leisure history. The second half of the chapter will explore modern historiography of leisure in the work of writers such as Dewey, Marx, Weber and Veblen, along with a review of their contemporary critics. In exploring this historiography, the question about the very *Modernist* ontological nature of leisure (as seen in the work of many leisure sociologists) will be addressed and shown to be epistemologically sufficient, but not necessary, for accounting for the unique nature of the relationship between work and leisure in the modern, Western world. Finally, two examples of twentieth-century leisure will be discussed, to highlight the communicative and instrumental diversity of modern leisure: long-distance walking in England and the invention of world music as a genre of pop music.

Defining modernity

The first concept we must look at is modernity. Again, we are moving away from the societal definition to a more fundamental question: what is modernist thought, and where did it come from? The societal definition of modernism equates the concept with the Fordist, brave new world of mass-production capitalist society (Bramham, 2006; Spracklen, 2009; Blackshaw, 2010). The big questions are tackled by Marx and Weber, and there is no doubt that Western society has entered a new era. Historically, the modern period begins with the Enlightenment, and the modernist paradigm is often described as the Enlightenment project. Just as (Western) society changed to a more urban industrial capitalist style, so, the argument runs, did our lines of thought. We became rational, as opposed to all who went before us, interested only in facts garnered via something called the scientific method, which in theory followed a hypothetico-deductive model of reasoning which bracketed out things such as values and religion and superstition. It was a project in search of the Truth, a project undertaken by men (middle-class white Western men) who were unattached to their project's resources and results. The sum of human knowledge was added to with every experiment, with theories tested and rejected or accepted through the scientific method. In philosophy, the scientific method was supported by positivists in the early twentieth century, giving rise to a term (positivism) that is often used alongside descriptions of modernist thought. This modernist project, then, was the project of Truth seeking that came to dominate Western society, and it is to this that the postmodernists are reacting.

Or so they claim. The description above raises many problems the postmodern literature has ignored. First of all, we must question this description itself. What is the historical credibility of the description? I will show that the modernist paradigm was a nice idea, but in historical (epistemological) reality, the project was never really all-embracing. This postmodern stereotype of the enemy is Whig history, a convenient story that has no resemblance to what occurred in the history of ideas. I will argue that not only does the modernist paradigm have origins further back in history, but so too does the postmodernist paradigm. Also, having discussed these origins, it will become clear that what really went on is something completely different to either of the paradigms. For instance, the Enlightenment project debated different ways to the truth; and science, the exemplar of modernist thought, does not follow the method it is said to have. Rather, the important epistemological site is that of rhetoric. I mentioned positivist philosophers, for example, but I did not say they were trying to rescue the Truth from Nietzsche, who had postulated the absence of truth in the late nineteenth century – the period when the modernist paradigm was supposedly at its peak.

The historical backdrop to postmodernism is then denied, creating a troubling paradox. For I originally wrote that postmodernism is a historically specific response to a project identified as modernism. Yet, if there was no modernism, how can postmodernism be an adaptation of modernist epistemology? The answer, of course, is that postmodernists believe in the ontological reality of modernism just as much as scientists believe in the scientific method, which is the method of modernism. So, in a sense, they obscure the epistemological history of the idea by resorting to a simplistic Whig history that they use to justify their own paradigm. We set out to try to rescue postmodernism and the quest for the Truth, which we have identified as existing before the mythical Enlightenment project. To do this, we must explore where these concepts of relativism and crude scientism came from.

We can show that the modernist paradigm was not subscribed to by its supposed founding fathers by taking many examples from the history of science. Isaac Newton was strongly religious and wrote apocalyptic commentaries on the Book of Daniel, from where he justified his theories on gravity as the force of God emanating through the Universe (Snobelen, 1999). He also spent more time studying alchemy than optics (Dobbs, 1982), and most his optical work he either plagiarized from others or proved to be true by the simple expedient of becoming head of the Royal Society and declaring that apparatus that did not give favourable results was flawed (White, 1998). A similar fixing of results occurred with the experiment that proved Einstein's special theory of (cosmological) relativity, when Eddington (a supporter of Einstein and the man charged with testing the theory) dismissed results that disproved the theory because the telescope used was 'unreliable' (Hudson, 2003). To go further back in history, Galileo's great crime was not saying the earth moved – which was nothing revolutionary – but satirizing the Pope and disagreeing with a Jesuit astronomer close to the Papal Court.

In more recent times, the validity of gravity waves and the Big Bang theory are all examples of scientific truths that are constructed, contested, and validated through methods far away from the hypothetico-deductive method (Collins and Pinch, 1993). Anthropological studies also show that scientists are as prone to shortcuts, deceptions and 'working backwards from the conclusion' as the rest of us (Latour, 1987). Throughout the previous three centuries, we can see natural philosophers renaming themselves scientists and creating a mystique about the True Path, the right way to truth unhindered by God or Parliament, whilst at the same time they do anything to keep their tenures and respectability by employing a whole host of rhetorical devices. I do not deny that many scientists try to be scientific, but because there is debate over what science is, and they are as culturally restricted as the rest of us, this interest in being scientific does not alter the non-scientific nature of the work they are involved in. This is because, like modernism, science is another rhetorical construction overused and contested even in this present time (the 'scientific' age – see Gooday, 2004). Science the method and theoretical ideal is not necessarily identical with science the practice. And just as the practice has changed, with what counts as science being an important question of the age, so too has the theoretical ideal undergone shifts in what it is and is not. To make matters worse for the scientific-modernistic historian, the term is a nineteenth-century invention, and before that there was no science, only epistemologies that influenced the dominance of science in the nineteenth century. The Natural Philosopher is continually confused with the Scientist, and present-centred misconceptions of the historically local idea and culture frameworks are lost (Ashplant and Wilson, 1988; Wilson and Ashplant, 1988).

There is a misconception running through all the literature that has now become clear. The modernist paradigm, which our erstwhile postmodernist is railing against, was not simply invented and used from the Enlightenment onwards. Whilst it would be fair to say that questions revolving around the rational, disinterested, objective pursuit of the beast called Truth were raised during the intellectual climate of the late eighteenth century, so were many other questions. And these questions that look like the scientific, modernist, rational cultural stereotype we are all aware of (in this scientific age we have all been enculturated from birth with the optimism of the white coats, though current trends in culture tend towards a grim suspicion of the white coats, though the power of them, for good or ill, remains) do not define an intellectual age. The modernist paradigm was not an invention of the Enlightenment. All we can say is that some philosophers followed epistemologies that resembled parts of the modernist stereotype, though these epistemologies were often competing with each other. One good example of this is the conflict between Priestley and Lavoisier (see Toulmin, 1957). The modernist stereotype adopted by contemporary thinkers who like to think they are modernist (and, ergo, the thing postmodernists react against) is backdated to these Enlightenment natural philosophers, who are called scientists and discoverers of oxygen. Lavoisier becomes the archetypal noble scientist, the rational man, who is sent to the guillotine by the (irrational) revolutionary mob. But Lavoisier and Priestley were competing against each other over who defined the nature of Nature. Priestley was a believer in the alchemical element of fire present in the air, called phlogiston, and his famous mice experiments proved its existence by discovering de-phlogisticated air, which was also enough evidence for God the Creator who made everything in the world in balance. Lavoisier was sponsored by the French government, which he worked for as a tax collector, and he believed in another kind of world, and another elemental design. He argued that phlogiston was not real – instead, there was an active substance he dubbed *oxygene*, the life-giver, which was the actual spirit of life. When he was executed, oxygen was not the thing we understand as oxygen now, and he was killed because of his connection with the old tax regime.

The historical basis of the modernistic epistemology is far more complex than the Whiggish histories imply. Throughout the history of ideas, there are rigorous, logically robust, arguments that run concurrently with each other, some being based around the search for some Truth (be it classical Aristotle, medieval Aristotelian scholasticism, Platonism, Arabic Aristotelianism, Neo-Platonism, hermeticism, Renaissance humanism, Renaissance scholasticism, Renaissance and so on) and others that remained sceptical (from classical scepticism through to Hume and empiricism, and ending with postmodernism). These epistemologies cannot, of course, be pigeonholed in this way, but the error is noted and necessary stylistically to explain the wealth of knowledge, knowledge seeking and philosophizing that has marked the human endeavour (especially since I have dealt only with the Western tradition, which pales into insignificance compared to Arabic or Chinese philosophy). It is also necessary to realize that these traditions brought with them a whole range of rhetorical stances inside their boundaries, so that conservatives, radicals, sceptics and those who were convinced of their humble path towards Truth could all be part of the epistemological paradigm. Even scholasticism, so long seen as a moribund, God-centred mediaeval enterprise, could and did allow a wide range of views within its framework, so that different commentaries on Aristotle, and translations from different sources, became the meeting points of intellectually opposed minds (Grant, 1997). Also, it is necessary to realize that the sequence from the golden age of classicism through to the barbarians of the Dark Ages, then to late classicism and scholasticism, followed by the progressive trend of truth in humanism, natural philosophy then science and modernity is ridiculous and anachronistic: the Dark Ages were not dark in terms of human achievement; scholasticism did not just die; there was never one all-embracing epistemological framework (there have always been hegemonic struggles within the history of ideas); philosophers who lived at the same time as others would turn in their graves if we assumed they worked with the same epistemological understandings; and we have already seen why modernity is an anachronistic, present-centred misnomer.

Milton, Gibbon, Mill and Habermas

Milton is the epitome of the seventeenth-century Protestant free-thinker and radical. His *Paradise Lost* (2003[1667]) re-tells the conventional story of Genesis and the Fall of Man with such perfect ambiguity that many of his later Romantic adherents believed the poet to be on the side of Satan (Danielson, 1999). Certainly, Satan gets the best lines in the poem, and his demands for freedom echo the rhetoric of Milton's *Areopagatica* and his other political works (Campbell and Corns, 2010). When Adam and Eve are banished from Eden, Milton again dangles the choice of Adam like the choice of every Englishman: to be free of tyranny, to have control over one's own destiny (Danielson, 1999; Durha and Pruitt, 2008; Cefalu, 2009; Campbell and Corns, 2010; Martin, 2010). It is a deliciously subversive poem, one that allows a multiplicity of readings (Danielson, 1999; Durha and Pruitt, 2008) and one which, for our purposes, expresses the view that there is and should be a difference between idle curiosity and worthy, self-organized leisure. Again, Milton makes the case in his other writings for a number of freedoms: of conscience, from despotism, to have time to do as one pleases. Milton catches the mood of other sixteenth-century radicals and puritans such as Richard Baxter, but he goes much further in terms of freedom and leisure (Cefalu, 2009; Martin, 2010). As Borsay (2005), Blackshaw (2010) and others (see discussion in Danielson, 1999) have argued, Milton's defence of private and public liberties formed a key turning point in modern notions of leisure time: freedom away from the obligations of work, where one could have time and space to pursue one's interests (provided those interests were not immoral or illegal). This led to an increasing importance of leisure in the public sphere, and a stronger demarcation of the private and the informal leisure activities of the domestic space (Habermas, 1989[1962]).

In the century after Milton, Gibbon stands out as someone looking back to the Renaissance and forward to the Enlightenment. He was a gentleman of leisure, from the new respectable middle class of England, who visited Rome as a tourist and pilgrim, where he was inspired by the ruins of the Roman Forum to write his masterpiece *The History of the Decline and Fall of the Roman Empire* (2005[1776–1788]). Writing this book itself was an endeavour of serious leisure (Stebbins, 1982, 2009), a slow and arduous journey of many years, reading primary sources and previous histories, as well as consulting books of numismatics and other reference works (O'Brien, 1997; Gossman, 2008). In addition to writing this book, Gibbon dabbled with the sports and habits of his class and gender (Porter, 1988) – and these enthusiasms shaped his view of the causal factors of the Roman Empire's decline. He believed that the onset of Christianity changed the character of the Roman ruling classes, diminishing their nobility, their love of wisdom and their desire to find good things in this life and not the next one (Porter, 1988; Gossman, 2008). Gibbon also believed that the empire became too autocratic, ruled by a series of soldier-emperors who allowed society to stagnate by snuffing out free will and free thought. The example of Diocletian, who resigned office to tend to his cabbages, was followed by the brutality of Constantine and his sons. The empire fell to the barbarians, who embraced individual courage and independence, and freedom to act according to circumstance. These barbarians hunted and rode horses, just like eighteenth-century gentlemen. But they stand for a particular version of individuality and liberty that Gibbon puts in a favourable light, next to the evil of absolute rule (Porter, 1988). This is Gibbon showing the way forward: the path to the Enlightenment follows the end of tyranny and autocracy, and the establishment of individual rights (O'Brien, 1997). Life, liberty and the pursuit of happiness (in one's leisure time) follos from this rejection of divine rights for human ones. Gibbon's work gave his readers a deathly account of the cloud of unreason that descended on Europe in the Middle Ages, but Gibbon's narrative lens, shaped by his musings at the Roman Forum, gave his readers the confidence to see they were Moderns, looking back at the Golden Age of Rome, but also the leaden age of medieval Christendom (Gossman, 2008).

In the nineteenth-century, as discussed in previous chapters, the philosophy of private pleasure and public restraint was associated with John Stuart Mill (1998[1859]). It was Mill who articulated the belief that the role of the state was to allow as many people as possible (men and women, a dissenting argument for his time) access to the public sphere, where they could contribute to political debates and take part in formal, public leisure and cultural activities. However, Mill also believed that the state had no role in policing the private lives of individuals. In private, one could do anything with one's leisure time; providing the choices of leisure activities did not harm anyone else (see Spracklen, 2009). Individual liberty was paramount for Mill in the private sphere – in public, a more virtuous role for individuals was assumed, which could result in individual freedoms being restricted for the common good. Mill, for example, would see complete sense in the public smoking bans enacted by many Western countries in the twenty-first century: he would support the ban on smoking in public if presented with the evidence of the dangers of passive smoking, though he would defend the right for individuals to smoke in private where their smoke did not drift into another person's breathing space.

Modern historiography of leisure (for example: Borsay, 2005: Rojek, 2005; Bramham, 2006; Blackshaw, 2010; Roberts, 2011) identifies these moments in the growth of Modernity as being essential to the construction of leisure. Certainly, leisure is something that only becomes meaningful and familiar to us in this period, when the focus on individuals, freedom and work allow people to see leisure as something universal and something to which all humans have a right (Borsay, 2005; Spracklen, 2009). Modern historiographers of leisure are correct to see this period as important: we cannot think of leisure, sport and tourism without thinking through the language game (Wittgenstein, 1968) provided by the advent of the modern world. We can see their ideas about the purpose of leisure provide a sustained engagement, following the Habermasian critical lens, with contemporary leisure history.

Habermas (1989[1962], 1990, 1992) himself writes extensively about this juncture. From discussing the work of the first theorists of Modernity, Habermas moves to articulating the consequences of the industrialization, secularization and individualization of society associated with the Modern turn (Habermas, 1989[1962], pp. 155–56). In analysing the role of Modernity in the decline of the family, he develops a model of leisure as a form of consumption in a state-capitalist political economy. Modernity, for Habermas, destroys the traditional role of the family and the notion of family property, family norms and family values. The enculturation of children through the family is lost, and with this loss comes a decline in the power of the family as a private institution. In compensation, politicians and society establish support systems for the family, such as welfare policies, public health and doles of various kinds. However, this support is not just material, for Habermas the Modern state intervenes in other words to provide structured mechanisms for managing the life of private individuals. There is a further erosion of the private, and the development of an instrumental network of consumption of many things, including leisure. As Habermas continued (Ibid., p. 156), 'the family now evolved even more into a consumer of income and leisure time, into the recipient of publicly guaranteed compensations and support services. Private autonomy was maintained not so much in functions of control as in functions of consumption; today it consists less in commodity owners' power to dispose than in the capacity to enjoy on the part of persons entitled to all sorts of services.' What is happening in Modernity is the transformation of traditional social structures into ones dictated by instrumentality: so the old, private space of the family is colonized by the rationality of commodification, and codes of honour are replaced by shopping lists; in turn, leisure becomes measured by the timing of television programmes. Habermas continues his strong critique of the instrumentality of Modernity and capitalism, and leisure behaviour becomes an ersatz copy of communicative, rational behaviour, a shallow, meaningless replacement for the public sphere of the Enlightenment. Habermas is not dismissing all leisure activity, only that which is imposed on people by those in power as a hegemonic trick to make the people forget about the power of public discourse and reason. With the shrinking of the public sphere, Habermas writes (Ibid., p. 159):

> private people withdrew from their socially controlled roles as property owners into the purely personal ones of their noncommittal use of leisure time.... Leisure behavior supplies the key to the floodlit privacy of the new sphere, to the externalization of what is declared to be the inner life. What today, as the domain of leisure, is set off from an occupational sphere that has become autonomous, has the tendency to take the place of that kind of public sphere that at one time was the point of reference for a subjectivity shaped in the bourgeois family's intimate sphere.

The theme of leisure behaviour continues, when Habermas contrasts the bourgeois culture of the Enlightenment with the empty leisure lives of us moderns, caught as we are in a realm of instrumental consumption. In making the contrast, Habermas also elucidates the distinction between affairs (action) self-directed by individuals pursuing private interests, and affairs (action) that unites individuals into a critical, communicative public sphere. In the Enlightenment, Western Europe was, for a brief moment, shaped by the latter: now, as in the last hundred and fifty years or so, caught as we are in a lifeworld colonized by instrumentality, our public sphere and our capability of free, communicative action, are limited. How we use our leisure time is indicative of our inability to choose freely or wisely; leisure defined as something apolitical or frivolous (something to do to keep us amused from boredom, as hundreds and thousands do every lunchtime when making friends on Facebook, perhaps) is itself evidence of the drain of critical thinking from the public sphere.

> So-called leisure behavior, once it had become part of the cycle of production and consumption, was already apolitical, if for no other

> reason than its incapacity to constitute a world emancipated from the immediate constraints of survival needs. When leisure was nothing but a complement to time spent on the job, it could be no more than a different arena for the pursuit of private business affairs that were not transformed into a public communication between private people. (Ibid., p. 160)

The early stages of Modernity, the rise of the nation-state and the homogenizing effects of capitalism and globalization, lead inexorably for Habermas into the present times, and our late modern dilemmas of freedom, free will and constraint. Here, Habermas is at his most pessimistic, and most in debt to his mentors from the Frankfurt School (see Spracklen, 2009). His strong criticism of the leisure activities of the culture consumers reads like a British leisure studies paper written in the 1980s by Bramham or Critcher: leisure is not only denuded of any political philosophy, but in its shallow, instrumental form it limits and constrains the ability of people to realize they are limited and constrained. As he continues (Ibid., p. 163): 'In the course of our [the twentieth] century, the bourgeois forms of sociability have found substitutes that have one tendency in common despite their regional and national diversity: abstinence from literary and political.' That is, the twentieth century has seen a decline in the engagement of people with critical, communicative debates about the nature and direction of politics, a decline equated by Habermas with the decline in discourses about the meaning and value of literature (and indeed other forms of high culture). What was crucial, for Habermas, about the critically-debating public in the Enlightenment public sphere, was its reliance on private reading and learning, which allowed for the intellectualization of the critical, public debate. In contemporary society, however, Habermas suggests (Ibid., p. 163), 'the leisure activities of the culture-consuming public, on the contrary, themselves take place within a social climate, and they do not require discussions.' There is in instrumentalized leisure and culture, an absence of both private cogitation and public discourse.

Leisure in historiography: Dewey, Marx, Weber and Veblen

John Dewey's *Democracy and Education* (1916) rests on his pragmatic philosophy, his belief in liberal democracy, and his insights into the psychological needs of humans for rational discourse and communication. He argued that leisure time is crucial for the development of humans, as it is only in such free time that humans can communicate with each other away from the constraints of work and toil. Drawing on the work of Milton and Baxter, Dewey identified the seventeenth century as the cradle of freedom. Dissenting arguments about religious and private liberty were then enshrined in the eighteenth century in the public sphere of the Enlightenment, and the Constitution of the United States. The important of free journalism, leisure time to read newspapers and debate political and social matters, follows from his belief in the life-enhancing nature of free thought that 'not only does social life demand teaching and learning for its own permanence, but the very process of living together educates. It enlarges and enlightens experience; it stimulates and enriches imagination; it creates responsibility for accuracy and vividness of statement and thought' (Dewey, 1916, original version available on-line at Project Gutenberg, http://www.gutenberg.org/files/852/852-h/852-h.htm). This, for Dewey, entails both the need to shape education in a way that develops children's ability to think critically, but also the minimization of constraints on adult's leisure lives. Dewey's account is seductive, and there are strong correlations between this account and the account of the public sphere and communicative rationality in the work of Habermas. The weakness, of course, is in the essentialism of Dewey's psychology of the mind and motivation, the perfect human finding solace in learning and choice (Dewey, 1933) – and the question of who controls education systems in the first place.

Marx's view of the progress of history is based on his dialectic of oppression and resistance (1992[1867]). For Marx, the key shift from feudalism to industrial capitalism in Europe occurs when the bourgeois classes accumulate wealth and thus social, cultural and political powers. The Enlightenment for Marx is the period when the bourgeoisie impose on the public sphere and universalize notions of privacy, liberty and individualism (see Marx and Engels, 2004[1848]). This, argues Marx, is a trick that hides the truth of their enslavement from the proletariat in the mills of the industrialized West. Leisure becomes bifurcated: into the polite cultural forms of the elite, which establish and limit status and taste; and the instrumentally useful rough sports, drinking and gambling of the masses. Following this historical materialism Marx predicts a series of revolutions culminating in a utopia of worker control and instructive, worthwhile leisure activities. This, as I have remarked earlier, is not something that has happened yet (Spracklen, 2009) – and despite the power of Marx's critique of capitalism and the obviousness of his arguments about class inequality, his history is reductive in its use of simple epochs, and specific to the conditions of Western Europe.

Weber provided a more sophisticated account of historical change leading to an account of life in Modernity (1992[1922], 2001[1930]). As discussed in the last chapter, Weber saw the growth of industrial capitalism as a result of earlier debates about the value of work, and the freedom of the individual, in the seventeenth-century Puritan communities of England and America (Weber, 2001[1930]). The rise of capitalism leads to the rationalization and globalization of the economy, which, in turn, leads to the rationalization of wider society and culture. Greater liberties for workers lead to more free time, but that free time is spent consuming products of modern factories and bureaucracies: the markets depend on the exchange of wages for match tickets, pies, holidaying by the seaside and similar factors. Modernity becomes identified with a particularly plastic form of leisure; one created by industry, consumed by the masses, then discarded for the next thing that comes along. We can see the modern sport, leisure and tourism industries as being the end-product of such instrumental calculations. Certainly, some things in the modern world and modern leisure sit easy with Weber's work, but there is little place for agency or resistance, and Weber's understanding of rationality seems to be predicated on the rational actor of economics, not the communicative rationality of the Enlightenment.

Veblen's *The Theory of the Leisure Class* (1970[1899]) critiques the emergence in Europe in the late eighteenth century of an elite class obsessed with fashions and material goods. This leisure class is the forerunner of the leisure class that exists in America and Europe in the late nineteenth century: the idle rich buying clothes and cars, the newly-established bourgeoisie climbing the social scale by joining the right clubs, and the desperation of those who feel the need to legitimate their belonging by conspicuously consuming goods. Veblen's caustic account of the leisure class is laced with a dry wit. Sports, in Veblen's (2005, p. 90) view, become

> expressions of a pecuniarily blameless life. It is by meeting… ulterior wastefulness and proximate purposefulness, that any given employment holds its place as a traditional and habitual mode of decorous recreation

Veblen's leisure class is the embodiment of late modern society. Although his historiography has been criticized for its historicizing bias (Diggins, 1999), the leisure class is instantly familiar to scholars and readers of the book today. We all know the inequities in our own social networks: the sons and daughters of capitalist families on year-long vacations; the middle classes showing off in their latest off-road vehicles or posing in their skiing gear on Facebook; and the wannabe bourgeoisie spending beyond their means on middlebrow restaurant experiences, weekend vacations in Prague, and five-hundred dollar walking-boots. In exploring this historiography, the question about the very *Modernist* ontological nature of leisure (as seen in the work of many leisure sociologists) can be seen – theories of leisure in the past simply become a transferral of meaning and purpose from the time the author was writing the book in question. In the work of Dewy, Marx, Weber and Veblen this does not detract from the accuracy and usefulness of their historiography of leisure. Hence, we can see that this modernist ontology of leisure can be taken to be epistemologically sufficient, but not necessary, for accounting for the unique nature of the relationship between work and leisure in the modern, Western world.

Leisure in the twentieth century: walking in England

For anyone who has pulled on a pair of walking boots at the foot of a Lakeland hill, Alfred Wainwright needs no introduction. His guidebooks on walks in the north of England remain in print and in use, and his public memory is increasingly managed in a respectful way by the work of the Wainwright Society (Wainwright, 1955, 1957, 1958, 1960, 1962, 1964). Wainwright's books were published at a key moment in English social history: when hill-walking and long-distance walking were still seen as part of a working-class social movement (Snape, 2004), but a time of increasing individualization and privatization of leisure practices (Borsay, 2005). Palmer and Brady (2007, p. 399) have argued that Wainwright's guidebooks establish a landscape aesthetic:

> in which certain upland features, particularly those of the English Lakes, define what is most valuable in landscape qualities. This aesthetic, based as it is in fell-walking, foregrounds the body and ways that valuing landscapes may be deeply rooted in practice.

Walking had become fashionable across classes (Borsay, 2005), and a combination of legislation and motorway-building made the Lakes more immediately accessible in the period. Pressure to develop long-distance walking in the early twentieth century came from the same working-class lobby as those seeking the 'right to roam' – those seeking the freedom to walk in the hills and open spaces of the country, which were almost all closed to ramblers by wealthy landowners. The struggle between those who wanted to enjoy the freedom of the hills, and the landowners who wanted to make profit from shooting rights, was a key moment in English working-class history. The ramblers who fought the landowners over right to roam also challenged those same landowners over access to hill-tops and wild-camping sites on long-distance walking holidays. The first call for a long-distance walk along the 'long green trail' up the Pennines was envisaged by Tom Stephenson: a rambler who campaigned for open access for all his adult life (Stephenson, 1989). In an article written in 1935, he set out his idea for a Pennine Trail after contemplating the problem of access (Stephenson, 1935, p. 17):

> When two American girls wrote asking advice about a tramping holiday in England, I wondered what they would think of our island ... Wherever they go, from Kent to Cornwall, from Sussex to the Solway, they will see these wooden liars; on the edge of many a tempting wood they will be confronted with the blatant warning. By the banks of luring rivers, on bare downlands and shaggy moors they will read 'Strictly Private.'

After the Second World War, the establishment in the United Kingdom of National Parks led to the recognition of rural spaces as tourist spaces (Merriman, 2005). This recognition, of course, also constructed rural tourism as an alternative to agriculture as an income-generator in rural economies and led to the idea of walking as a holiday and walking as leisure. Long-distance walking holidays became a *legitimate* tourist pursuit, bringing walkers into National Parks and other rural areas: whether areas of some recognizable aesthetic quality desired by walkers, or areas in which local authorities tried to capitalize on the pursuit of long-distance walking by attracting walkers to revitalize economies marginalized by the effects of post-industrialism (Morrow, 2005; den Breejen, 2007; Bold and Gillespie, 2009).

In 1938, Wainwright decided to walk up the Pennines, to Hadrian's Wall, and back again. He wrote a book about his experience, but this was left unpublished until 1986. It is not clear when exactly Wainwright wrote the book, but in the foreword to the published book, Wainwright claims to have written it over a number of months immediately following his return. Certainly, the book captures the fears and rumours of the impending war. The book is reputedly published exactly as it was first written. In the foreword, Wainwright (1986, p. viii) writes:

> I put it [the manuscript] away in a drawer where it lay forgotten until quite recently when ... My publisher asked to see it. I dug it out of

> hiding and brushed off the dust...Not a word has been changed, not a word omitted or inserted.

In this book, the framework of tourism support services for walkers is absent. Wainwright only has Ordnance Survey maps to guide him from village to village, and he stays in pubs and hotels where the clientele are local farm labourers, or passing salesmen. The paths he walks take him through farmyards and out onto moors where the paths fade away. The book describes a different landscape to the bucolic world of the Lakeland guidebooks. There are moments when he is happy, of course, but the tone of the book is cold and miserable. Instead of the ideal landscape aesthetics of those books, *A Pennine Journey* describes featureless hills, rough villages, dark woods, brown moors. For example, on the landscape (pp. 63–64):

> The aspect, seen under a leaden sky, was drab and uninviting in the extreme...There is no suggestion of warmth or comfort.

The dull, uninviting nature of the land is reflected in the sullen and unfriendly locals he meets on his travels. The farmer Rowley is described in great detail as a frightening ogre, feral and threatening. On the customers in one pub, for example, he observes, in a moment of *petit bourgeois* revulsion (p. 172):

> They were all big, rough fellows from the farms...their presence made me feel thoroughly out of place.

A Pennine Journey clearly influenced Wainwright's view of the Pennines. The part of the Pennine Way he seems to enjoy the most in the *Pennine Way Companion*, the section along Hadrian's Wall, was the inspiration and destination of his 1938 journey. Indeed, in the Introduction to the *Pennine Way Companion* (Wainwright, 1968, p. ii), he argues that the route of the Pennine Way is wrong to go through the Cheviots to Scotland, and should have 'as its thrilling finale, the exciting arrival at the Roman Wall!' This is the obvious connection between the journey of 1938 and his 1968 Pictorial Guide to the Pennine Way. But it also quite apparent some of the dark landscape aesthetic of that 1938 journey is reproduced in Wainwright's notes for the 1968 guidebook. So, in both books, there is a sense of boredom, of long trudges through an uninspiring, monotonous land, and of unfriendly natives; in the personal notes in conclusion, Wainwright (1968, p. xxii) returns to the theme of gruff farmers and the menace of farmyards he mentions so often in *A Pennine Journey*. Thirty years of tourist industry development have not changed Wainwright's view: the Wall is nice and historical, but he would rather be across to the west, where his beloved Lakeland mountains are in view, in sight of walkers on the Pennine Way on the top of Cross Fell, 'superbly arrayed across the wide valley of the Eden' (Wainwright, 1968, p. 56). The happy coincidence of the name of the River Eden, with its connotation of paradise, seems to be too good for Wainwright to miss. Lakeland is paradise, and Wainwright, on top of Cross Fell, starting glumly at its tempting peaks, feels like he has been thrown out of it.

Wainwright used his writing to construct notions of taste and distinction (Bourdieu, 1986) in walking, which, in turn, limited – sociologically and geographically – the impact and commercialization of long-distance walking holidays in the Pennines, compared to the Lakes. He established a dark imagined community (Anderson, 1983) through the aesthetics of those north and north-eastern hills, and the people who live in those hills, in juxtaposition to the green and pleasant land (and people) of the Lakes. And he did this partly because, as someone from Lancashire, he carried about with him a notion of Yorkshireness (more strictly, 'Pennineness') that reinforces, and is reinforced by, stereotypes about the people and the place.

World music, or the instrumentalization of instruments

Pop music emerged in the twentieth-century as a successor to the ragtime, jazz and blues that were situated in African-American popular culture. Such forms were commodified through appropriation and exploitation by white producers and white-owned record companies (Oliver, 1990). The invention of recording technology, radios and record players had allowed, in the early twentieth-century the global dissemination and consumption of professionally-constructed music (Bennett, 2001): this allowed music industries to be created in many countries producing recorded music based on a mix of local and global forms (for example, the *Rebetika* of Greece, see Holst, 1977). The limitations of the industry's technologies, and the demands of radio stations that needed space for advertising, led to a format of recordings of short duration, typically three minutes (Oliver, 1990). Pop music came to prominence in the second half of the twentieth century, when white entrepreneurs and radio disc jockeys took the rock-and-roll form of blues to white America, with white poster-boy Elvis Presley (Guralnick, 1995). With this successful formula – songs with hooks, easy choruses, screaming girls, good-looking young men and women – rock-and-roll became the first global pop music (Dettmar, 1997). Young people across the globe listened to the music, admired the ways of America, and formed their own bands. Rock-and-roll itself became rock, while pop and dance and a number of different genres established themselves in the industry, the media and the record shops (Bennett, 2001).

World music is part of the wider globalized popular music industry, with its own websites, magazines, record labels, festivals, managers, booking agents and bands. Its definition is, of course, not strictly policed, but it has come to mean 'Roots' music from local/national/regional cultures including the West; Global fusions and dance music; and pop and rap music from various local/national/regional cultures beyond the West. World music was invented in London in the 1980s as a term to embrace the genres above, while rejecting Western pop music, as well as any form of rock (Bohlman, 2002). The consumption of world music, and the genre itself, is primarily Western in orientation. Musicologists have argued that world music is a product of white, Western hegemony, or more generously, a response to globalization, hybridity and diaspora (for example, Connell and Gibson, 2004). Others, such as Corn (2010, on Yothu Yindi), argue that some forms of world music are rooted in genuinely local cultures, which serve to define communities and belonging. Skinner (2010, p. 36) on Amadou and Mariam, argues:

> *Dimanche à Bamako* is a salutary example of a collaborative musical recording that transcends this anxious/celebratory dichotomy. It is an 'Amadou and Mariam' album produced 'by and with Manu Chao'; a collaboration that bridges the Global divide between North and South by artfully weaving the civil and wild experiences of African communities, from Bamako to Paris.

As first sketched out in the work of MacCannell (1973, 1976), authenticity, something real or essential in a place or experience, was the ultimate goal of every tourist. The morality of authenticity elides smoothly into a Western, middle-class sensibility of culture: the authentic is good because it runs counter to the homogenizing tendencies of globalization, because it encourages diversity and respect and cultural heterogeneity. MacCannell sees authentic cultures as existing, in a *Gofmannesque* sense, backstage. Although authenticity and the quest for it have played an important part in the research agenda for leisure (for example, Rojek and Urry, 1997; Wang, 1999; Aitchison, 2006; Matheson, 2008;

Andriotis, 2009), the concept has been the subject of much academic criticism and development. The authentic in world music is where we can see Habermasian communicative reason at work (Habermas, 1984[1981], 1987[1981]), in the agency of individuals attempting to challenge the restrictions of the music industry. Hybridity here is the key to a critical understanding of the role of world music in the construction of multiple identities, but such construction, while demonstrating the agency presupposed by Brah (1996), Jacobson (1997), Solomos (1998) and others, is limited by the instrumentalized structures of Western society and the whiteness of Western national identities. Modern leisure remains fundamentally instrumental in character.

11 Future Histories of Leisure

Finally, this chapter begins with an examination of contemporary leisure history, with a discussion of postmodern history (Zagorin, 1999). I will explore the way in which postmodernism and postmodernity have opened up opportunities for communicative discourse over the meaning and purpose of leisure: particularly through the rise of the Internet and popular music. However, the commodification of sport and the malign influence of trans-national corporations on the choices of leisure consumers will serve as a reminder that contemporary leisure's history is a Habermasian history of communicative retreat against the rising tide of instrumental rationality. In the second section of this chapter, I will explore the way in which dreams of communicative leisure fulfilment have been, and are, expressed through the depiction of leisure in alternative histories, such as science-fiction and the fantasy stories of J.R.R. Tolkien (see Fimi, 2009). I will argue that such alternative histories of leisure can be seen as ways in which free, rational choices about good and bad are made free from persuasion; and ways in which we can avoid the wrong, instrumental choices. However, there is an irony in the way these alternative histories are created by the industry of capitalism to meet a superficial, consumer demand.

Contemporary leisure history

Although Veblen (1970[1899]) was the first sociologist to notice changing patterns and significations of leisure consumption associated with the increasing affluence of Western elites in the first half of the twentieth century, it was only in the period following the Second World War that leisure (in the modern West) started to be associated with the construction of cultural identities. Some scholars observed that young people used their leisure lives and spaces to create alternative, counter-cultural identities (Hebdige, 1979). Stanley Parker (1972, 1976) noted the emergence of leisure consumers, and leisure choices made within the flux of rapid societal change; Ken Roberts argued that increasing concern with leisure in a post-industrializing West was leading to the establishment of leisure policies and managers in the public sector, and leisure industries in the private sector (Roberts, 1978). Other sociologists and philosophers started to argue that changes to working practices brought about by automation, computerization and globalization would result in more free time for individuals, and hence more need for leisure activities (Smigel, 1963). Rojek describes this thesis (2010) as the *Leisure Society Thesis*, which he strongly critiques as being naïvely utopian about the value of leisure as a freedom, and the meaning of free-time. He associates the thesis with positivist American leisure sciences in the second half of the twentieth century, which, he correctly argues, blithely assumed more free time, and therefore more worthwhile leisure, would be the consequence and the moral good of Western modernity. It is not given that changes to Western society will lead to that utopian world of free, limitless leisure. Further, it is not clear that leisure in the Leisure Society, as Smigel understood it (1963), would necessarily have any moral purpose. As will be seen in the section below on science-fiction, it is possible to imagine a leisure society where the leisure lives of those who belong to the society are amoral or immoral.

The consumption of leisure and the role of leisure in identity building led to a strong feminist attack on the work of Parker and Roberts; sociologists of leisure argued that free choices and free time for leisure pursuits and identity formation were not options for marginalized women forced into private, domestic roles (Deem, 1986, 1999; Aitchison, 2000). Similar structural critiques of the liberal theory of leisure arose, making the same point the feminists made: social groups such as the working classes (Clarke and Critcher, 1985; Coalter, 1998) or the poor in the developing world, or minority ethnic groups in the West, did not have the power or the freedom to choose and partake fully in leisure activities (Bramham, 2006).

More recently, it has been argued that leisure changed so much at the end of the last century, and the beginning of this one, that a simple definition eludes our inquiry (Rojek, 2005, 2010; Blackshaw, 2010). Youth sub-cultures have proliferated with the globalized reach of Western popular music forms, and in playing with these sub-cultural forms, individuals have been transgressing modern structures (Bennett, 2001; Rojek, 2002). With the rise of the Internet, there is no way of knowing the historic specificity of any leisure practice or personal avatar beyond a claim on a web-site, or a shared set of imagined histories (McGuigan,

2006). There is no history of contemporary of leisure, as such, only a set of narratives with no objective arbiter of epistemological value. In *The Meaning and Purpose of Leisure*, I argue:

> The paradox of leisure – as freedom, as constraint – is one that both approaches to leisure theory recognize. But all previous attempts to escape the paradox have collapsed into one discourse or the other. Leisure studies, it seems, have found themselves in an epistemological crisis. This crisis surfaced in the belated recognition of structural changes to society and culture underpinned by a shift to post-Fordist, post-industrial, postmodernity (Lyotard, 1975:1984). In response, many leisure researchers have abandoned theory for empiricism or specialism. Others, such as Rojek... have attempted to re-engage with critical sociology and shape ideas about epistemological doubt, the multiplicity of truths, and the fracturing of social structures to establish a theory of postmodern leisure, or postmodern leisure theories. While the postmodern turn has been useful in understanding the complexity and the fragility of meaning, and has brought to leisure studies more awareness of consumption, postmodernism itself has been the subject of a sustained, theoretical critique within leisure studies (Bramham, 2006). In concluding the discussion on the paradox of leisure, I return to the normative question of change at the heart of critical theory: if we are to remain committed to challenging inequality, we must remain committed to some notion of truth and justice. (Spracklen, 2009, p. 3)

It is argued that leisure has become postmodern or has been changed due to the shift towards postmodernity evidenced in Western societies (Rojek, 2010; Blackshaw, 2010). Leisure has become, as Blackshaw (2010) argues, following Bauman (2000), liquid. That is, it the nature of the subject of our inquiry (leisure) has changed as society has become postmodern. In my own research on rugby league, for example, the pessimism and nostalgic yearning of Hunslet and Bramley supporters contrast with the excitement and optimism initially associated with the commodified spectacle of Rupert Murdoch's new world order of rugby league (Spracklen, 2009; Spracklen et al., 2010). When News Corporation first offered millions of pounds of funding in exchange for exclusive rights to televise rugby league in 'Europe' (in practice, England) and Australia, the game of rugby league was split between loyalists (traditionalists) and those who took the cash or bought the global (expansionist) vision (Kelner, 1996). In the north of England, some rugby league fans opted to defend the game against any change or expansion (Kelner, 1996; Spracklen, 1996). But ultimately, in England and in Australia, the game's clubs, sponsors and administrators all accepted and welcomed the involvement of Murdoch's global media empire (Denham, 2004; Collins, 2006). The growth of television coverage, via News Corporation (which owns the Sky satellite channel), of the European (English) Super League and the (Australian) National Rugby League has led to the game expanding into new regions and territories around the world (Spracklen, 2007; Spracklen and Spracklen, 2008). In many of the developing areas, the involvement of students, attracted by television coverage, has been a catalyst to drive expansion (Collins, 2006). All the respondents from the international rugby league on-line forum, by their very involvement in the game, were believers in the globalization of rugby league: they all saw the benefits of television exposure and commercial sponsorship, and all welcomed the professionalization of the game (Spracklen and Fletcher, 2010). However, despite their enthusiasm for expansion and globalization, they were reticent about abandoning rugby league's working-class history. Most of the respondents had some connection to the game's working-class heartlands of Brisbane, Sydney or the north of England: either they themselves were born there, or they had family there, or they had lived there for some period. As such, they saw in their expansionist work the project of a spreading the idea of rugby league as a working-class masculine game: even when they admitted that they themselves were not working class or they explained that rugby league in their particular developing country was played by middle-class students.

Clearly, rugby league is globalizing and, in doing so, demonstrates all the material, demographic, technological, social and cultural flows of globalization (Appadurai, 1995; Holton, 2008); as well as the dominant flow of westernization (commmodification, professionalization, Americanization via Australia) identified by Hall (1993), Giddens (1990) and Bauman (2000). Leeds Rhinos exemplify this globalizing rugby league phenomenon, and Headingley Stadium is an outpost of the conspicuous commodification and consumption of professional sport (Horne, 2006). Watching the World Club Challenge between the Rhinos and the Storm on television, with the sound of the commentators muted, one could be forgiven for mistaking the match for any other global team-sport spectacle: the prominence of sponsor logos, the huge crowd in the dazzling lights of the stadium, the multicultural nature of the hypermasculine professional athletes, the designs of the jerseys and the tricks of the TV studio, all part of the Americanization of global sport (Denham, 2004). Turn the commentary up, however, and despite the Americanized style of delivery, the rough tones and flattened vowels remind the listener of rugby league's connection to the imaginary, and the imagined, working-class world of the north of England.

There is, then, one contradiction. Rugby league is, like its union counterpart, a commodified product, its elite competitions part of the global calendar of passive consumption (Horne, 2006), its international profile fuelling participative and commercial expansion into new markets. In this expansion and development, the postmodern nature of commodified sport becomes apparent (Maguire, 2005). News Corporation and other multinational sponsors create supranational leagues, and rugby league clubs change and lose their local identity and become businesses in the same way as elite football clubs have done (Fawbert, 2005). As Denham (2004) has argued, rugby league's embrace of Americanization and commodification, a postmodern turn itself, is evidence for the dissolution of identity (reference here) and liquidity of structure (Bauman, 2000) associated with postmodernity. The existence of Leeds Rhinos and the World Club Challenge provides evidence for some postmodern shift away from high modernity (Giddens, 1990), from the traditional, fixed working-class communities and identities typified by the Hunslet of Richard Hoggart (1957). Leeds itself is a microcosm of the move towards postmodernity and the decline of the traditional, industrial base of the working-class economy. The globalization of rugby league demonstrates some embrace of the rationality of instrumental capitalism at the end of modernity (Habermas, 1990), perhaps the beginning of postmodernity (Harvey, 1989; McGuigan, 2006). But in Hunslet and Bramley, individuals still choose to identify with working-class communities of modernity, expressing communicative action (Habermas, 1990) in resisting the conformity of Super League. What rugby league in Leeds demonstrates is the complex nature of globalization, the relationships among the local, national and global that Featherstone (1990) argues is the crucial tension in global, globalizing culture. Rugby league in Leeds appears to be an example of glocalization (Roche, 2006), the local adaptation and response to a globalizing trend (Giulianotti and Robertson, 2007).

What this demonstrates is that leisure is a place for fluid shifts of identity and intentionality (Blackshaw, 2010; Rojek, 2010): but at the same time, its liquid nature is solidified by our ontological, epistemological and moral frameworks. This is where Habermas helps us understand the shift towards intentionality and liquidity. Faced with the commodification of leisure, and its instrumentalism, we can accept the consequences of limitation imposed on us. Or we can use leisure to find meaning and purpose in our everyday lives in a world where traditions, cultures, communities and histories are being constantly challenged. Of course, in choosing to find some sense in leisure and the history of leisure, we have to meet the challenge of postmodernism in the subject field of history – how can we write a history of leisure, or even a history of contemporary debates about the meaning and purpose of leisure, if we have no way of finding an objective arbiter of truth?

Postmodern history

As Evans (1997) and Jordanova (2000) have argued, historians have a duty to try to write history that is sensitive to the subjective nature of sources and the act of writing, but also faithful to the values of the discipline, such as trust, honesty and reliability. Can this really be the answer to what constitutes good history? Their defences of history are, in part, a response to others such as Zagorin (1999), Jenkins (1991; 1995; 1997), Joyce (1995) and Ankersmit (1989), who have all argued that the postmodern turn – the claim that all history is subjective, that there is no objective standard of truth – has led to the death of history, or at least history that claims to reveal some truth about the past. The postmodern enterprise in history is designed specifically to improve our historical knowledge, to remove ethno- and present-centred discourses, and to stop judicial and Whiggish versions of the past. What *really* happened, if anything really real happened at all, is the goal, though because our knowledge and language are inextricably connected to the present, such an enterprise becomes forlorn. Yet there is a progressivist trend within history, provided by postmodern readings, that allows the past to speak for itself without the present imposing its interests on it (and by exploring possible interpretations within the discourse of the past, we lessen the impact of our own present-centred interpretive mechanisms). And it is precisely this progressive nature of historical discourses that implies there is any history at all to deal with. So whilst it is true to say the past has no ontological reality, it does have epistemological reality. And in the postmodern enterprise, the latter is far more important than the former. The problem with most postmodernist texts is that the writers are coming into the problem of the relativity of truth without understanding just where any relativistic, sceptical ideas came from. The postmodern view of truth is that there is no such thing, or rather there can be a number of competing truths with equal validity, differing only in the type of discourse and completely incommensurable. Alongside this realization of relativity and the language-boundness of any inquiry into any truth, comes a rejection of the rational, scientific method. Grand claims of identifying truth, reality, progressions and so on become just another form of rhetoric, and the Enlightenment project is cast aside.

There is, however, an intrinsic problem with all texts that support a postmodern epistemology. By rejecting any commensurability, by casting doubt on the claims of any epistemological framework that has ultimate answers, it makes itself invalid. The relativistic approach is the victim of its own success – by setting out a correct framework based on internal axioms (that is, the suggestion that there is no true way of doing things), it criticizes itself for doing the very thing it tries to proscribe. In a vicious circle, relativity ends up cutting its own throat.

Alternatively, the postmodern text will suggest that, instead of there being no Truth, distinct from language and epistemological frameworks, there is instead a multitude of truths, many paths to find personal enlightenment. It suggests, in other words, that we, as our own reality creators, make our own worlds and systems of truth, and that you can not compare one with another. The truths of the Catholic Church become as real as those of Taoism, and there is no way of saying, 'This one is better' or 'The other has a lesser truth value'. If this is the case, then there is nothing to stop the errant philosopher choosing the rational, scientific way instead of the relativistic anarchism that evolves from the work of Nietzsche, Kuhn and Wittgenstein to Feyerabend and Derrida (Fuller, 1992). If choice between truths is limited by interests, either in the form of cultural norms or personal drives, there is still the problem of the errant philosopher deciding to have an interest in being rational and scientific. (I use science and scientific throughout in the sense of the stereotypical view of the form of language-game that has diffused through Western academia, though I do not suggest this stereotype is a true representation of what the people who claim to be scientists actually do.)

The rhetoric of this section has, so far, used the term 'postmodernism' in an ahistorical manner, as if ideas including the idea of postmodernism somehow exist outside a historical framework within the history of human thought. Yet, the very term 'postmodernism' suggests a historicity of approach. It is beyond modernist thought. It is a network of related ideas that is a response to, or has grown out of, the modernist paradigm. It has a defined, cultural context of its own and has, in this essay, a very specific meaning. The word is overused in contemporary academia and has become a byword for rejections of mainstream thought. If there is a modernist approach to the truth, then it must be identified before we can criticize from a post-modernist position. There is a problem with most of the rhetoric identified under the label post-modernism, in that the conclusions of the postmodernist idea are given without any understanding of how involved in the pre-postmodern postmodernity actually is.

Even worse for the would-be philosopher is the hijacking, by some academics, of a very specific, temporally placed concept about the relativity of truth and the uncertainness of all knowledge. Now we are regaled with the term 'postmodernism' from every angle. Far from being an insight into the dangers of Whiggish history and judgemental truth (a concept that suggests we can never be sure of the Truth, and there is no adjudicating framework to say, for example, that the past is worse than the present which is best, or Einsteinian physics describes how the Universe actually is, or Western science is more rational and true than Islamic faith) postmodernism has become a keyword, a semiotic token that delineates a new paradigm separate from the old one of empirical hypothesis testing, which believes in the role of the researcher in interpreting research, in the value-ladenness of all ideas and actions, of the multiplicity of meanings and the incommensurability of paradigms. It is a fracturing of knowledge into discourses, which are inevitably influenced by language.

There is a historical dimension to questions about epistemology that often goes unseen by both the supporters and detractors of the post-modern enterprise. It is a product of its own age, yet it has come out, via Nietzsche, of a long struggle amongst philosophers over what there is to know, and how we can know what we know, if we know anything at all. I contend that the argument for a relativistic epistemology comes from a historical dilemma within the rational enterprise of philosophy, which was dealing with a concept that lay beyond the epistemological framework they were working in. It comes from within Platonic philosophy, and the concepts of *episteme* and *doxa*, and how these related to the dichotomy between the everyday, real world of man and the really real Forms which man could not ever hope to access other than indirectly. From this, the relativistic logical framework can be considered as working in a meta-logic that is situated in the same site as Plato's forms, and hence becomes an episteme epistemology, one that relates to a meta-logic and meta-truth, as opposed to a *doxa* epistemology, one that I think works within the framework of meta-truth and is the kind of relativistic truth which postmodernists have identified (Fuller, 1992). This solution to the problem of postmodernism allows, then, the response by Evans (1997) and Jordanova (2000): it is the practice of historians, the community of historians, and the respectful nature of both, that allow historians to write history, and for histories that have been through the rigour of peer-review to be judged to be of a better quality than those written on a random web-site.

So it becomes possible to write a history and philosophy of leisure that discusses contemporary debates about the meaning and purpose of leisure, one that is sensitive to historical objectivity, theoretical rigour and the shifting sands of both modern society and leisure theory. In light of the playfulness of the postmodern turn and the demands of its protagonists to seek truths outside the academy (Derrida, 1982; Eco, 1986), I will apply the debates about *The Leisure Society* and postmodern, liquid leisure, to a discussion of alternative histories of leisure in (post)modern, popular culture. This is not, of course, to argue that narratives by story-tellers have a truth-equivalence to books written by sociologists or historians. But it is useful to see how those writers see leisure – in the future, or in their alternative universes – as this helps us understand what they believe to be the meaning and purpose of leisure in the time and place they were writing.

Utopias, dystopias

When thinking about the future, science-fiction writers can imagine a utopia, a place where life and society is good. In the work of Iain M Banks (1988, 1989, 1992, 1997, 1999, 2010), for instance, the citizens of the Culture, a vast, intergalactic assemblage of planets and spaceships, have complete freedom to act in ways that they wish, with freedom from interference (Berlin, 2002). Leisure choices are free, and everyone gets to make good choices for themselves: the enormous wealth of the Culture, and its intellectual capital (tied up in the Artificial Intelligences of the Culture's millions of spaceships), ensures such liberties (for example, see Banks, 1997). Utopias are warming fairy-tales. In science-fiction the faith in the future resembles the scientism of the Victorian Age (Gooday, 2004) and the belief in progress associated with the American theorists of the 'leisure society' (Smigel, 1963; Rojek, 2010): technology will free us from work to live enriching lives of leisure. Utopias are popular in middlebrow science-fiction (Moylan, 2000), but utopian visions are not the whole story: science-fiction writers can imagine a dystopia, where the world has gone wrong. The implications for people's leisure choices of such dystopias vary among stories, but in essence leisure becomes something immoral or amoral, or something that is limited, constrained or otherwise controlled.

By the late nineteenth and early twentieth centuries, American and European beliefs in scientific and moral progress outlined in Chapter Nine were being challenged by an alternative view of change. Western society was becoming weaker: the Boer War saw a large number of recruits rejected on the grounds of ill-health (and this led to the growth of the Physical Education movement – see Soloway, 1982). Working-class labour movements were becoming more radical. And The Great War demonstrated that 'civilized' countries were perfectly capable of applying science and capitalism to slaughter on an industrial scale (Hobsbawm, 1995). These trends influenced the development of modern science-fiction. In the last decade of the nineteenth and in first half of the twentieth century, three novelists wrote books that tried to predict what the future (of western society, and, indirectly, of course, of leisure) would look like.

In H.G. Wells' *The Time Machine* (2005[1895]), a time traveller from 1895 arrives in the future to find people called the Eloi. These seem to live in some sort of paradise of continual leisure. But they have no art, no creativity and no intelligence: the leisured classes have devolved and forgotten science and technology. The Eloi's society is kept going by the Morlocks: bestial, ape-like cannibals treated as slaves by the Eloi. Aldous Huxley's *Brave New World* (2007[1932]) is set in London in 2540. There is no war or poverty, and people seem to be happy. But all people seem to do with their lives is have sex and take drugs (specifically, a substance Huxley calls *soma*). Like the Eloi of *The Time Machine*, the people of this brave new world have no care or understanding of art, culture, literature, science or philosophy. George Orwell's *Nineteen Eighty-four* (2008[1949]) is also set in London, in the year of the title. London is part of Oceania, which is controlled by a totalitarian, Stalinist Party, which, in turn, is controlled by Big Brother, a dictator whose presence is felt in every aspect of citizens' lives. Society in Oceania is divided between members of the Party – who are subject to 24-hour surveillance and control – and the Proletariat, who live in squalor and are fed propaganda. Members of the Party are forced into morally good leisure activities; the Proles get to drink cheap beer and use pornography created by machines, the ugly stuff of everyday consumption.

By the end of the twentieth century, dystopias increasingly appear in popular science-fiction. The best example is the twenty-second century Earth portrayed in the long-running *Judge Dredd* strip in the British comic 2000 AD (1977-ongoing), created by John Wagner (Barker and Brooks, 1998). Dredd is a Judge of Mega-City One, a city of eight hundred million inhabitants on the east coast of America (reduced to four-hundred million by various major wars in the years the strip has been running). The world is postnuclear, and the Mega City is a fascist police state controlled by judges like Dredd, who act as judge, jury and executioner. There is no democracy, but there is unfettered capitalism. Most of the citizens of the city are unemployed, as robots do most of the jobs. But living in the city is better than living beyond the walls in the irradiated landscape of the Cursed Earth. In this dysfunctional future, rampant consumerism dictates the passing fashions and passive leisure lives of the Mega City's bored citizens: stories from the comic strip include trends in cosmetic surgery to make people ugly ('Otto Sump's Ugly Clinic', first appearing in 2000AD 186, 15 November 1980); aggro domes where citizens can vent their frustrations against robots ('The Aggro Dome', 2000AD 183, 25 October 1980); and illegal 'fatty' competitions, where professional gluttons eat off against each other until one competitor is left alive ('Requiem for a Heavyweight', 2000AD 331–334, 27 August 1983 to 17 September 1983).

Adventures in the holodeck

The science-fiction television series *Star Trek*, created by Gene Roddenberry as a way of showing 1960s America some kind of liberal-democratic utopia, has spawned a multibillion-dollar industry (Tulloch and Jenkins, 1995; Bernardi, 1997; Weldes, 1999; Hark, 2008). It is one of the entertainment industry's global brands, with the original TV series and five spin-offs created in the 1980s and 1990s (*The Original Series*; *Star Trek: The Next Generation*; *Deep Space Nine*; *Voyager*; and *Enterprise*) watched in hundreds of countries around the world (Weldes, 1999; Hark, 2008). The original TV series (TOS) reflected the optimism of 1960s America, with the Captain of the USS *Enterprise*, James T. Kirk, embodying the youth and hope of the recently assassinated John F. Kennedy. The crew of the *Enterprise* were multicultural, multinational and, with the presence of half-Vulcan First Officer Spock, multispecies (Bernardi, 1997). Kirk and the crew fight Klingons, destroy computers that control Orwellian societies, and support the careful scientific exploration of the galaxy (Hark, 2008). *Star Trek* is set in a future where Earth is one of a number of worlds in a United Federation of Planets. As imagined by Roddenberry and the writers of TOS, the Federation is a liberal, utopian society: there is no anti-social disorder, no environmental problems and no capitalism (indeed, no money), and everybody has equal rights, opportunities and outcomes. The message is positive: everyone gets along in the Federation, and they bring democracy and liberal ideas to the rest of the Galaxy. The dramatic tension of the series comes from the new life and new civilizations that Kirk and his crew come across in their five-year mission, boldly going where no man has gone before.

In the 1980s, the franchise was revived with a new version of the ship and a new crew, boldly going 75 years further into the future than Kirk's adventures. *Star Trek: The Next Generation* (TNG) was again created by Gene Roddenberry, and again the Federation was portrayed as a perfect utopia (Hark, 2008). However, as the series progressed and Roddenberry relinquished editorial control, TNG started to explore problems within the Federation, and the personal relationships of those on the crew. The series reflected some of the obsessions of 1980s America: accumulation of wealth, search for identity, spiritual satisfaction and sensitive leadership (typified by Captain Jean-Luc Picard's insistence on having meetings with his senior staff whenever something threatening appeared in front of the ship).

One of the noticeable changes between the *Enterprise NCC-1701* captained by Kirk, and Picard's *Enterprise NCC-1701-D*, is the concern in the new ship with its crew's wellbeing. For Kirk and his gang of adventurers, time off-duty was spent in a cramped recreation room, eating artificially-coloured processed foods, playing 3-D chess or cards. Occasionally, there would be singing or a Shakespeare play, if they happened across a troupe of actors on the final frontier, and in a couple of programmes there was an opportunity for 'shore leave'. Casual sex with beautiful women was, of course, always on the cards for Kirk (playing the Kennedy archetype), but for the rest of the crew there were strict, quasi-military rules about how they behaved. On the *Enterprise* of TNG, however, the pressures of work were recognized. The senior officers included a Ship's Counsellor, the crew were encouraged to bring their families with them into space, there were regular concerts and there was even a cocktail bar that resembled something out of a five-star hotel. Instead of food rations, each cabin had a replicator, a machine that could create any food or drink you desired. But above all that, there was the holodeck: an area of the ship that used 'matter transport' technology to create any environment you wanted to explore, including simulations of anyone you wanted to meet. The technology created the illusion that the holodeck stretched far beyond the bounds of the ship, allowing the crew's android Data, for example, to walk the streets of Victorian London as Sherlock Holmes. With the holodeck, leisure opportunities were literally infinite (even if, with tedious regularity, it broke down and created a dramatic narrative situation).

In *Deep Space Nine* (DS9), the universe of the Federation meets a number of different alien societies at a space station (DS9) in the Bajoran system. On the space station is a bar run by an alien called Quark, who, as a Ferengi, values profit above all else and is not averse to breaking rules to make that profit. He takes cash in the form of gold-pressed latinum from the Federation officers and others on the station – the moneyless society of the Federation stops in the frontier culture of Bajor. As well as Quark's bar, there are shops on the promenade selling Bajoran takeaway snacks and other Bajoran consumer goods, as well as Klingon restaurant (where one can eat fresh *gagh*, or living serpent worms) and an upmarket tailor's run by a Cardassian. The promenade, in other words, resembles some kind of off-world bazaar, where characters from the series shop, relax and hang out. In Quark's bar, there are *dabo* tables, where scantily-clad hostesses take bets from clients and waiters serve enough *synthehol* drinks to make sure the gamblers do not notice that the gaming tables are rigged. Upstairs, Quark has holosuites for paying customers, which include a wide array of erotic programmes: that said, the Federation officers who use the holosuites mainly use the technology for outdoor sports, detective games and other, more *earnest* leisure activities, as if they all have read the work of Stebbins (2009).

Individuals have the freedom (liberty) to explore their own identities, to be leisured in their lifestyles (to use their own agency), but the postmodern free-for-all is absent – people make moral choices. In *Star Trek*'s imagined future, leisure is initially bounded by the morality of liberal democracy. Kirk believes in an American ideal of freedom, where every individual must be liberated from structures to be allowed to be themselves (Hark, 2008): it is no surprise that the series reserves its biggest anger at the totalitarianism of Romulans and Klingons, or the worlds where computers dictate every moment of individual lives. In 'The Apple', for example, Kirk even destroys a computer (the ultimate Gramscian metaphor) that has made a planet safe for its inhabitants, because in making it safe and full of small pleasures, the computer has denied the inhabitants the chance to make any choices of their own. But Kirk's Federation is tied also to an American morality, which dictates that with freedom comes a responsibility to make correct choices (Barnardi, 1997. This morality is seen in the leisure choices of the Federation officers across *Star Trek*'s films and television programmes. Officers on starships are expected to have the cultural capital to appreciate classical music, to have a desire for climbing and canoeing, as well as a curiosity for learning. They are like the philosopher-kings of Plato's *Republic*, virtuous, wise, intensely moral, but also strong, resourceful and psychologically balanced. Their use of leisure and the holodeck/suite is as a moral amplifier to their labour, not merely something to help them unwind.

Worlds of fantasy

Fantasy fiction as a discrete literary genre emerged in the twentieth century, though, of course, it had significant antecedents in earlier literature. What made fantasy fiction an accepted part of the book industry (if not the canon of elite taste) was the phenomenal success of some of its pioneers: Lord Dunsany, C.S. Lewis and the most important (in terms of influence on the genre, and total sales), J.R.R. Tolkien. The work of anthropologists such as Douglas (1991) and Levi-Strauss (1963) stresses the mythological power of blood and the strength of blood-ties in tribal units. Historically and in the present day, Western European culture puts a metaphysical emphasis on blood and heritage: whether it is the powers of our Royalty, the new nobility of celebrity, our obsession with family roots or the chimera of race. 'Blood' is an old folk belief, one that has both universalist tendencies in human mythology, and one that is tied through Anderson's notion of the imagined community (Anderson, 1993) to nation-states, (re)constructed ethnicities and self-serving histories (Hobsbawm, 1992). The folk myth of blood and race/nation is easily observed in the work of Tolkien, who was deliberately evoking these notions at a time when England was in the middle of a debate about Englishness (Easthope, 1999). In *The Lord of the Rings*, the main human character, Aragorn, is a hero and king-in-exile because he has the blood of the higher men, the Numenoreans, running through his veins (Tolkien, 1954/55). His views on taxation, war and peace, governance, diplomacy and other issues are irrelevant to his claims to the throne of Gondor: when we read the book, or watch the film, we rarely doubt his pure-blooded status, its implications of a supreme race with a holy right to rule the world. Aragorn's claim to the throne is all in the blood – now, of course, we would say it is all in the genes.

Tolkien's work evoked a world called Middle-earth, which was originally meant to be an earlier version of our world, one set in a distant time before the shape of the land was changed, and the ascent of Man led to the rise of what we know as history (Shippey, 2005). The world is the site of a Manichean struggle between light and darkness, good and evil – between the angelic Valar of the blessed land of the West, and the Satanic Morgoth and his lieutenant Sauron. This struggle was a 'secondary world' expression of Tolkien's deeply-held Catholic faith: Illuvatar in the work of Tolkien was a supreme God who is Tolkien's Catholic God. Tolkien populated Middle-earth with trolls, dwarves, hobbits, humans, orcs, goblins, ents and elves. These latter beings were immortal and bound to fade in power or retreat to the West as the cultures of the other species (or races, as they are commonly called in Tolkien's work and commentaries on Tolkien) become dominant. First, the elves are sorely beaten by Morgoth and his orcs, as recounted in the epic cycle of the First Age published in *The Silmarillion* (Tolkien, 1977). At the end of *The Silmarillion,* there are struggles with the dwarves, who are corrupted by lust for elven treasure in the halls of King Thingol. In the Second and Third Ages, elves have moments of long peace but they lack the will to flourish beyond a few pockets of cultural excellence: the forest realm of Lothlorien and Imladris, home of Elrond Half-elven. The Third Age is the Age of the Numenoreans, who build the great kingdoms of Arnor and Gondor. It is the Numenoreans who allow the hobbits to settle in the land called The Shire. Gondor itself has its own empire, its own martial history and its own slow decay – it is the Byzantium of Tolkien's world (Shippey, 2005).

The evolutionary history of Tolkien's writing has been told elsewhere (Shippey, 20005; Fimi, 2009). What is relevant for this history and philosophy of leisure is the framework of story-telling that appears in the first iteration of Tolkien's writing. In *The Book of Lost Tales,* written between 1916 and the early 1920s (Tolkien, 1983, 1984), a mariner called Eriol travels in some numinous sense to the Cottage of Lost Play in Tol Eressea, the elven isle. There he is entertained as a guest in a medieval banquet: he drinks and eats, listens to music, then, at the end of the evening, he listens to stories of elven history. This is everyday life among the elves of the isle, who are exiled from the lands from whence Eriol has travelled (lands that were evidently the lands of our world). The elves of Tol Eressea, on first reflection, live fulfilled leisure lives. But they are trapped by their history and their tragedy. Whenever they come together, the old stories are told of their struggle and loss. Story-telling is an essential part of how these elves construct their identities; it gives them meaning, and their leisure, purpose.

Tolkien's first published novel, *The Hobbit* (1937), grew from a story he made up for his children, and is a book deliberately written for children (Shippey, 2005). It has a playful narrative voice and a knowing, arch style quite unlike the seriousness of *The Lord of the Rings*. However, it does draw on Tolkien's mythological writing, directly through the appearance of Elrond and the mention of Gondolin, and indirectly through the appearance of elves, dwarves, goblins and the barren Northern European ecology of the wild lands through which the hobbit travels (Fimi, 2009). In the story, Bilbo Baggins is introduced as a peace-loving hobbit of the Shire, a country loosely based on the Mercian landscape of Tolkien's childhood: a *merrie* England of farms, winding lanes, tall trees, and taverns. Hobbits are never portrayed playing cricket, but they partake in recognizably rural English leisure pastimes. They smoke pipes, eat cakes, sup flagons of ale, go on country walks, attend fairs and enjoy fireworks: in *The Hobbit*, Tolkien even makes a bad joke about how the game of golf was invented by a hobbit knocking off the head of a goblin king called Golfimbul, which then landed in a rabbit hole. In *The Lord of the Rings*, the more grown-up sequel to *The Hobbit*, the bucolic idyll of the Shire is expounded on in more detail, but the essentials remain the same. For the hobbit men, the hard day's graft in garden or field is rewarded by a drink of beer or cider at *The Green Dragon*, where songs are sung, fiddles are played, and one expects (but never sees) a traditional English skittles alley. Of course, hard graft is not viewed as something hobbitish at all: in the Shire, the hobbit men and children see nothing wrong in skipping from work to play in the woods. Hobbit men who are 'respectable', that is, those of the middle and upper classes with inherited fortunes, spend their time tending their gardens, hunting or reading comforting books. Hobbit women are less carefully observed (a problem with all of the women in the books, with the exception, perhaps, of Galadriel of Lothlorien): their roles are strictly delineated as domestic, child-caring or, in the case of Lobelia, scheming widows.

At the beginning of *The Hobbit*, Bilbo Baggins finds himself unexpectedly hosting an ever-increasing party of dwarves, who make outrageous requests for a diverse and particular range of foodstuffs and beverages. Fortunately, Bilbo manages to satisfy their demands, as he has a well-stocked larder and the cultural capital to play a respectable host. In *The Lord of the Rings*, the story starts with Bilbo hosting his birthday party, which involves the purchase of food, drink, musicians, presents and, of course, fireworks made by Gandalf the wizard. Consumption is a key to social relations in the Shire. Like some real-life *potlach* tribute, Bilbo's parties show he is wealthy enough to allow his guests to consume as much as they want. As well as parties, the other most obvious hobbitish leisure activity in *The Hobbit* and *The Lord of the Rings* is the inn or tavern. I have already mentioned the pub in the Shire. But the pub is also central to Bree, a town halfway between the Shire and Rivendell, where Frodo Baggins and his companions meet Aragorn the Ranger at *The Prancing Pony* inn. When Frodo returns to the Shire, after his epic quest through distant lands, it is to Bree that he comes first.

Hobbit parties and pubs are juxtaposed with the leisure lives of others Frodo and the others meet on their quest. In Imladris and Lothlorien, the sad elven songs are sung, and tales told of things that happened thousands of years ago. In Rohan, Tolkien recreates Anglo-Saxon manners and mores in the mead-hall of Meduseld (Fimi, 2009). In Gondor, in the city of Minas Tirith, meals are taken quietly, and the Steward Denethor spends his free time delving into half-forgotten secrets that leads him to find the *palantir* seeing-stone. This magical stone is linked to one captured by Sauron, and Denethor is ensnared by evil, twisted by his own quest for the lore of elder days. It is no surprise that the hobbits return to the Shire, where they can smoke pipes, drink ale and clap and sing, away from the horrors of decaying cities, or the feudal demands of the kings of Rohan and Gondor.

Middle-earth's ideal leisure life, the hobbit comforts of pipe, pub and good company, were only the comforts of Tolkien himself (Carpenter, 2002). In his published *Letters*, Tolkien makes constant reference to the tedium of academic teaching and management (Carpenter, 2006). The freedom to explore his research topics (around old Northern European languages) was, like with any other academic, restricted by the instrumentality of university life. Tedious meetings and arguments about policies drained his creative and intellectual powers. What kept him going was the knowledge that he could nip down to the pub for a few pints with his old chums, where he could smoke without anyone complaining, where he could be away from work and his wife, and where he could recite some of his own poems or read some of his stories. The intentionality (Rojek, 2010) of this is clear enough: the construction of his self-identity, his need for male company, his need for an escape from the pressures of work and domesticity.

So as it was for Tolkien, so it became for the hobbits. And so it became for the fantasy genre, because *The Lord of the Rings* established a set of genre rules for others to follow (see Gray, 2010). Fantasy fiction had to include elves, or immortal creatures that were bittersweet about the past and had some liminality about themselves, and also orcs, or other-named bad things with a poor education and propensity for violence. Fantasy fiction had to be a struggle between good and evil, and the battle for good had to be won by someone simple-hearted. And fantasy fiction portrayed worlds where ale was quaffed by the flagon, pipes were smoked and characters spent their evenings telling stories and singing songs in inns that were a pastiche of early twentieth-century English pubs and Chaucerian taverns. The stereotypes of fantasy fiction were established: even those who tried to reject them were influenced by them in rejecting them (Shefrin, 2004; Gray, 2010). The fantasy inn became popularized in late twentieth-century role-playing games such as *Dungeons and Dragons* and *Runequest,* which were the pre-Internet versions of multi-player fantasy games (Toles-Patkin, 1986). Hence, the perfect evening of an Oxford professor became a natural part of the globalized, gaming cyber-culture of *World of Warcraft* (Chen, 2009), and a teenage boy from South Korea could find it perfectly normal to key in a request for another foaming tankard of ale from the busty barmaid on his computer screen.

12
Conclusions

If there is a brief conclusion that captures this book's argument in one phrase, it is this: leisure is a fundamental part of the human condition, but our freedom to choose our leisure has always been constrained. In this final chapter, I need to summarize the endeavour of the previous ten chapters, return to the histories of leisure discussed in some of the other chapters, and identify future research projects within the framework of philosophical history, and historical philosophy, of leisure. Before I do any of this, it is important to situate this book in current debates in leisure studies. As an academic discipline, the field of leisure studies continues to produce engaging reflections on the meaning and purpose of leisure. Three books demonstrate the vitality and depth of leisure studies, as well as the diversity (Bramham and Wagg, 2010; Crouch, 2010; Rojek, 2010). Before I summarize and conclude, it is worthwhile to spend some time discussing each of these books and what relationship they have with this book.

Chris Rojek is one of a handful of social theorists on leisure who have consistently produced provocative and intellectually stimulating work. In *The Labour of Leisure*, Rojek (2010) continues the two related themes of his previous book *Leisure Theory* (Rojek, 2005). The first theme is that academic work on the meaning and purpose of leisure has been undertaken on a false promise that leisure is something simplistically associated with freedoms of choice and time. That is an overly confident sleight-of-hand. There is no reason to doubt that there are still very important and influential leisure researchers – especially in the United States – who see leisure as something merely voluntary, something merely associated with what we do when we clock off work (see Roberts, 2011, for an excellent summary). But most researchers and theorists of leisure would refute Rojek's argument that their work on leisure is a naïve hangover from some positivist, utopian paradigm.

The second theme is that Rojek is offering a way of reconciling the concerns about leisure raised by the structuralists in the 1980s and 1990s with the poststructural turn to identity expressed in postmodernist accounts of leisure. In the 1990s, of course, Rojek himself was one of the first theorists of leisure to critique structural theories of leisure through the lens of postmodernism (Rojek, 1995). In this new book, as in *Leisure Theory* from 2005, Rojek provides a way forward for leisure researchers and students. It was the snappy and energetic Action Theory in 2005; in 2010, he presents us with the woefully-expressed SCCASMIL framework, an improvement on the previous leisure research paradigm of SCCA (State-Corporate-Consumer-Academic) taking into account Social Movements and Illegal Leisure (hence, the ASMIL). The SCCASMIL framework is a necessary consequence of Rojek's novel embrace of emotional intelligence and intentionality as the key to understanding leisure. (Rojek is quite dismissive of simplistic models from social psychology, with the exception of emotional intelligence, which he introduces without any critique or any recognition of the strong arguments against the concept.) Rojek argues that SCCA is a weak framework for leisure studies, but with added SMIL the framework becomes a kind of power-assisted exo-skeleton that moves our focus of study 'from simple causal models of leisure choices and trajectories of behaviour to more complex perspectives that approach leisure experience as the product of relations between multiple equilibria' (Rojek, 2010, p. 188).

Such opaque writing, combined with the pomposity of the claims about SCCASMIL's uniqueness (as if no leisure researcher has ever considered social movements and illegal leisure before!), do not make for a clear and reasoned contribution to leisure theory, which leads to another major problem with his book: its intended audience. Rojek refers to students throughout the book, and he even writes Wiki-style definitions in separate text-boxes for learners whenever a big word like 'paradigm' is first mentioned. This is all very good, for a book aimed at students. But if the book is for students, then the jargon-heavy nature of the writing must be improved. If it not for students, then the text-boxes become patronising.

That said, Rojek is always an important voice in leisure theory. His account of the development of leisure studies is clear and interesting. His insistence that positivists and neo-liberals need to recognize the struggle over the meaning of leisure is sound; likewise, his concern with the reductive nature of structuralist arguments is still relevant. And his claim that leisure involves the management of intentionality, while not novel, is one that is corroborated by many other researchers on the borders between leisure and culture. Like his previous work on leisure, *The Labour of Leisure* is still clearly provocative and intellectually stimulating, albeit in a rhetorical sense.

David Crouch (2010) has produced a more reflective monograph than Rojek's, which draws on growing intellectual trends in wider critical circles. In literary culture in the West, journeys have become a familiar *trope* in which fractured identities, dissonances and emotional responses to space replace the old-fashioned plots and quests typified by the fantasies of Tolkien. These journeys are not necessarily the dalliances of a *flaneur* roughing it in the squalid underbelly of a modern metropolis, though there is some element of such voyeurism in all these journeys. Rather, books by authors such as W.G. Sebald draw us, as readers, into their meditations on the end of modernity, and the loss of innocence associated with the human condition. This is psychogeography, the realization that the space and place around us is not merely where we live, but where we shape our sense of belonging, our meaning and purpose, and where we live in a perpetual tension between our ability to impose meaning and the instrumental discourses that try to shape such spaces for their own ends. Just as the old saying goes that art imitates life, so in this book David Crouch joins the cultural geographers and anthropologists who imitate art imitating life by writing academic meditations on journeys, spaces and the individuals who make meaning of them.

This is not a postmodern play that scorns the traditional tools of academia. Crouch draws heavily on a range of theory, relying on theorists as different as Deleuze, Bauman and Urry, and utilizing familiar theories of leisure and tourism along the way. His approach to theory is wisely pragmatic. This book could have been a long exercise in reflecting on reflection on reflection on reflection of reflection, pinned to the flag of Derrida or Foucault. But Crouch avoids this by using theory when it is needed and allowing his own argument, his own research, to dominate the journey. He shows the reader that space is in a constant flux as people make their own way through their lives, and the act of creativity is fundamental to the construction of fleetingly real, solid identities and senses of space.

The third book is a collection of research papers edited by Pete Bramham and Steve Wagg (2010), *The New Politics of Leisure and Pleasure*. In the book, the editors attempt to demonstrate a clear link between leisure and pleasure – leisure in the early part of the twenty-first century, they argue, has become part of a wider politics of identity, having 'some kind of relationship to the politics of pleasure' (ibid., p. 269). This new cultural politics gives Western forms of leisure in the twenty-first century something of a postmodern, or poststructural, sheen. The editors show, through the choice of case studies in the substantive chapters, that the old politics of class and gender is (in the commodified, corporatized West, at least) being replaced by more fluid identity politics (what Blackshaw (2010) associates with liquid leisure).

This fluidity over identity is reflected in the fluidity of the concept of leisure used to define the content of the book. Sport, recreation and tourism do get chapters of their own, but these sit alongside chapters that would not look out of place in edited collections from cultural studies: Attwood (2010), for instance, writes about erotic play and its prevalence in what she calls the new leisure culture: Petley (2010) examines pornography on television; Drake and Haynes (2010) write about the deregulation of television; Blackman (2010) explores drug use and risk; and Wagg (2010) brings an analysis of comedy production and consumption to the collection. Clearly, Western society (and politics) has changed since the twentieth century, and the great projects of modernity and leisure have changed with society. As the editors write in their introductory chapter (Bramham and Wagg, 2010, p. 5):

> The leisure project of modernity, orchestrated by local and national policy-makers in pursuit of rational recreation has been superseded by new agendas, policy alliances and corporate forces... Alongside this there has been a pattern of deregulation in many nation states and growing globalisation experienced through the Internet, travel and new patterns of communication. In numerous cases what was once marginal and deviant has now become the basis for widely available commercial entertainment.

All three of these books identify typologies and discourses of leisure that I have loosely interpreted in previous chapters as being late modern: leisure associated with the specific time and place of the West at the end of modernity. Strictly speaking, it could be argued that such late modernity is equivalent to postmodernity, coming as it does after the grand era of modernity that spanned the years from the Enlightenment to the second half of the twentieth century. This is the view of some of Bramham and Wagg's (2010) contributors, and, hesitantly, Bramham (2010) himself: Western culture has changed so much due to the pressures of commerce and technological transformations that – for some of the new elite, at least – social structures have dissolved in a wave of liquid, hybrid and fractured identities. Leisure is a symptom or effect of this dissolution. Rojek (2010) and Crouch (2010) would agree with Bramham's assessment of leisure in this late or post- or liquid modern state, though they would differ about the nature and extent of the fluidity of identity and leisure in the particular moment of the early-twenty-first century West.

I agree entirely with these authors that leisure for the middle classes of the West has become associated with the sociology and politics of identity. It is clear from this book that I have identified such a shift in the modern history of leisure in the late twentieth century: at least for those rich or powerful enough to be able to buy such identity formation. However, such a shift is not necessarily a paradigm shift in a Kuhnian sense of the word. Contemporary leisure is not completely different from leisure at any other time – there are strong continuities of choice and of constraint in the meaning and purpose of leisure. What I have shown is that the history and philosophy of leisure is a history of communicative agency and instrumental control. Individuals from pre-historical times onwards have tried to find solace and expression in the time and space afforded to them, all the while finding other, stronger individuals using the bounds of tradition (of religion, of tribe, of government, of patriarchy, of hegemony, of feudalism, of heredity, of economic inequality) to keep others in their place. These traditions and assumptions are based in wider social and cultural contexts – uniquely contextualized in specific places and ages – but are used to define the limits of leisure through the centuries. Leisure has been the chain of slavery as well as the opportunity for self-realization. There is nothing new in the relationship between leisure and identity: every generation of every culture has made themselves anew through the things they did when they were not working, and every generation of ruler has tried to impose their authority through the instrumental use and limiting of leisure practices.

There are enormous dangers in trying to write a synthesis of this kind, and I am aware that the strength of my argument is limited by the lack of depth in the details. There is a need for a wider research project into specific histories of leisure in specific moments in time and space. The strength of histories of leisure that already exists in the history of modern sport and the social history of leisure in Britain needs to be emulated by other histories. Someone needs to respond to this book by fleshing out my fleeting sections – about nineteenth-century Japan, or late Byzantium, for instance – into full-length monographs. All such projects, of course, would be improved immeasurably by being framed by the philosophical reasoning about the meaning and purpose of leisure that I draw from the work of Habermas here and elsewhere. While the study of sport has been served excellently by the establishment of the discipline of the philosophy of sport (with academic associations, conferences, and journals), there is no corresponding discipline for leisure – apart from multidisciplinary leisure journals (*Leisure Studies, Leisure Sciences, Leisure/Loisir, Annals of Leisure Research*), there is no specific peer-reviewed journal for the philosophy of leisure and there is no association uniquely charged with promoting the philosophy of leisure.

In the introduction, I wrote that this book is intended to be a sequel to *The Meaning and Philosophy of Leisure*. Where that first book focussed on leisure at the end of modernity, and the effect of globalization and postmodernity on leisure, this book looks back at the meaning and purpose of leisure in the past. It is a history and philosophy of leisure. I argued that it is not enough to write a history of leisure on its own – in fact, it is impossible without engaging in the debate about what counts as leisure (in the present and in the past). I claimed that it is necessary to examine leisure and theories of leisure in historiography, critically and through the lens of philosophy. This book's aims, then, were twofold: firstly, to engage with academic debates about leisure in history and philosophy, which will lead to a strong critique of the narrow focus of previous historiography and social theory; and secondly, to provide a much broader chronological and geographical scope for problematizing leisure, which allows for both a more balanced analysis of the meaning and purpose of leisure, and a comparative exposition of that meaning and purpose in context. These aims have been met. I have discussed the philosophy of leisure across time and across different parts of the globe. I have problematized the meaning and purpose of leisure in the writing of leisure scholars and historians, both past and present. I have shown the importance of leisure through a number of historical eras, in places across the globe. I have shown that each historical context has its unique issues and debates, but that there are sufficient similarities across those contexts to establish a universal philosophy of leisure. That universal philosophy is ultimately associated with agency and with identity: leisure is something that makes us human.

References

Aitchison, C. (2000) 'Poststructural Feminist Theories for Representing Others: A Response to the Crisis in the Leisure Studies "Discourse"', *Leisure Studies*, 19, 127–44.

Aitchison, C. (2006) 'The Critical and the Cultural: Explaining the Divergent Paths of Leisure Studies and Tourism Studies', *Leisure Studies*, 25, 417–22.

Alcock, S., Cherry, J. and Elsner, J. (2001) *Pausanias: Travel and Memory in Roman Greece* (Oxford: Oxford University Press).

Allison, G. (2004) *Japan's Postwar History* (Ithaca, New York: Cornell University Press).

Almond, R. (2003) *Medieval Hunting* (Stroud: Sutton).

Ammianus Marcellinus (2004) *The Later Roman Empire*, translated by A. Wallace-Hadrill (Harmondsworth: Penguin).

Anderson, B. (1983) *Imagined Communities* (London: Verso).

Andriotis, K. (2009) 'Sacred Site Experience: A Phenomenological Study', *Annals of Tourism Research*, 36, 64–84.

Ankersmit, F. (1989) 'Historiography and Postmodernism', *History and Theory*, 28, 137–53.

Ansary, T. (2010) *Destiny Disrupted: A History of the World through Islamic Eyes* (New York: Public Affairs).

Appadurai, A. (1995) *Modernity at Large: Cultural Dimensions of Globalization* (Minneapolis: University of Minnesota Press).

Aquinas, T. (2003) *Selected Writings*, translated by R. McInerny (Harmondsworth: Penguin).

Arditi, J. (1998) *A Genealogy of Manners: Transformations of Social Relations in France and England from the Fourteenth to the Eighteenth Century* (Chicago: University of Chicago Press).

Armstrong, K. (2001) *Mohammed: A Biography of the Prophet* (London: Phoenix).

Ashplant, T. and Wilson, A. (1988) 'Present-Centred History and the Problem of Historical Knowledge', *The Historical Journal*, 21, 253–74.

Asser (2004) *Alfred the Great: Asser's Life of King Alfred*, translated by S. Keynes (Harmondsworth: Penguin).

Attwood, F. (2010) 'Sex and the Citizens: Erotic Play and the New Leisure Culture', in P. Bramham and S. Wagg (eds) *The New Politics of Leisure and Pleasure* (Basingstoke: Palgrave Macmillan).

Augustine (2002) *Confessions*, translated by R.S. Pine-Coffin (Harmondsworth: Penguin).

Babinger, F. (1992) *Mehmed the Conqueror and His Time* (Princeton: Princeton University Press).

Bacon, F. (2008) *The Major Works* (Oxford: Oxford University Press).

Bacon, W. (1997) 'The Rise of the German and the Demise of the English Spa Industry: A Critical Analysis of Business Success and Failure', *Leisure Studies*, 16, 173–87.

Baker, D. (1980) 'From Plowing to Penitence: Piers Plowman and Fourteenth-Century Theology', *Speculum*, 55, 715–25.
Bakhle, J. (2005) *Two Men and Music: Nationalism and the Making of an Indian Classical Tradition* (Oxford: Oxford University Press).
Balme, M. (1984) 'Attitudes to Work and Leisure in Ancient Greece', *Greece and Rome*, 31, 140–52.
Balsdon, J. (2004) *Life and Leisure in Ancient Rome* (London: Phoenix).
Balter, M. (2005) *The Goddess and the Bull: Catalhoyuk – An Archaeological Journey to the Dawn of Civilization* (London: Simon and Schuster).
Banks, I.M. (1988) *Consider Phlebas* (London: Orbit).
Banks, I.M. (1989) *The Player of Games* (London: Orbit).
Banks, I.M. (1992) *Use of Weapons* (London: Orbit).
Banks, I.M. (1997) *Excession* (London: Orbit).
Banks, I.M. (1999) *Inversions* (London: Orbit).
Banks, I.M. (2010) *Surface Detail* (London: Orbit).
Barber, C. (2002) *Figure and Likeness: On the Limits of Representation in Byzantine Iconoclasm* (Princeton: Princeton University Press).
Bard, K. (2007) *An Introduction to the Archaeology of Ancient Egypt* (New York: Wiley-Blackwell).
Barker, C. (1955) *Henry George* (Oxford: Oxford University Press).
Barker, M. and Brooks, K. (1998) *Knowing Audiences: Judge Dredd* (Luton: University of Luton Press).
Barkey, K. (2008) *Empire of Difference: The Ottomans in Comparative Perspective* (Cambridge: Cambridge University Press).
Barme, G. (1999) *In the Red: On Contemporary Chinese Culture* (New York: Columbia University Press).
Barnard, A. (2000) *History and Theory in Anthropology* (Cambridge: Cambridge University Press).
Barnes, B. (1977) *Interests and the Growth of Knowledge* (London: Routledge).
Barnes, T. (1998) *Ammianus Marcellinus and the Representation of Historical Reality* (Ithaca, New York: Cornell University Press).
Barstow, A. (1988) 'On Studying Witchcraft as Women's History: A Historiography of the European Witch Persecutions', *Journal of Feminist Studies in Religion*, 4, 7–19.
Bate, J. (2009) *Soul of the Age: The Life, Mind and World of William Shakespeare* (Harmondsworth: Penguin).
Baudrillard, J. (1988) *Selected Writings* (Cambridge: Polity).
Bauman, Z, (2000) *Liquid Modernity* (Cambridge: Polity).
Baycroft, T. (1998) *Nationalism in Europe 1789–1945* (Cambridge: Cambridge University Press).
Bayly, C. (1988) *Indian Society and the Making of the British Empire* (Cambridge: Cambridge University Press).
Beard, M. and Henderson, J. (2000) *Classics: A Very Short Introduction* (Oxford: Oxford University Press).
Beaver, D. (2008) *Hunting and the Politics of Violence before the English Civil War* (Cambridge: Cambridge University Press).
Beiser, F. (2005) *Hegel* (London: Routledge).
Bender, B. (1978) 'Gatherer-Hunter to Farmer: A Social Perspective', *World Archaeology*, 10, 204–22.

Benedict, B. (2002) *Curiosity: A Cultural History of Early Modern Inquiry* (Chicago: University of Chicago Press).
Benedikz, B. (2007) *The Varangians of Byzantium* (Cambridge: Cambridge University Press).
Bennett, A. (2001) *Cultures of Popular Music* (Buckingham: Open University Press).
Bentham, J. (1996[1799]) *An Introduction to the Principles of Morals and Legislation* (Oxford: Oxford University Press).
Bentley, M. (2006) 'Past and "Presence": Re-visiting Historical Ontology', *History and Theory*, 45, 349–61.
Berg, M. and Power, E. (1997) *Medieval Women* (Cambridge: Cambridge University Press).
Berlin, I. (2000) *The Roots of Romanticism* (London: Pimlico).
Berlin, I. (2002) *Liberty* (Oxford: Oxford University Press).
Bernardi, D. (1997) 'Star Trek in the 1960s: Liberal-Humanism and the Production of Race', *Science Fiction Studies*, 24, 209–25.
Berry, J. (2007) *The Complete Pompeii* (London: Thames and Hudson).
Birley, A. (1993) *Marcus Aurelius: A Biography* (London: Routledge).
Birley, A. (2000) *Hadrian: The Restless Emperor* (London: Routledge).
Birley, D. (2003) *A Social History of English Cricket* (London: Aurum).
Black, J. (1985) *The British and the Grand Tour* (London: Routledge).
Blackman, S. (2010) 'Rituals of Intoxication: Young People, Drugs, Risk and Leisure', in P. Bramham and S. Wagg (eds) *The New Politics of Leisure and Pleasure* (Basingstoke: Palgrave Macmillan).
Blackshaw, T. (2010) *Leisure* (London: Routledge).
Boas, F. (1988[1928]) *Anthropology and Modern Life* (Mineola: Dover Publications).
Bohlman, P. (2002) *World Music: A Very Short Introduction* (Oxford: Oxford University Press).
Boitani, P. and Mann, J. (2004) *The Cambridge Companion to Chaucer* (Cambridge: Cambridge University Press).
Bold, V. and Gillespie, S. (2009) 'The Southern Upland Way: Exploring Landscape and Culture', *International Journal of Heritage Studies*, 15, 245–57.
Bolton, R. (1976) 'Andean Coca Chewing: A Metabolic Perspective', *American Anthropologist*, 78, 630–34.
Bordo, S. (1987) *The Flight to Objectivity: Essays in Cartesianism and Culture* (Albany, New York: SUNY Press).
Borsay, P. (2005) *A History of Leisure* (Basingstoke: Palgrave Macmillan).
Bose, S. and Jalal, A. (2003) *Modern South Asia: History, Culture, Political Economy* (London: Routledge).
Bourdieu, P. (1986) *Distinction* (London: Routledge).
Bowen, H. (2008) *The Business of Empire: The East India Company and Imperial Britain, 1756–1833* (Cambridge: Cambridge University Press).
Bowerstock, G. (1997) *Julian the Apostate* (Cambridge: Harvard University Press).
Bowler, R. (1989) *The Invention of Progress* (Oxford: Blackwell).
Bradley, R. (1998) *The Significance of Monuments: On the Shaping of Human Experience in Neolithic and Bronze Age Europe* (London: Routledge).
Brah, A. (1996) *Cartographies of the Diaspora* (London: Routledge).
Bramham, P. (2006) 'Hard and Disappearing Work: Making Sense of the Leisure Project', *Leisure Studies*, 25, 379–90.

Bramham, P. (2010) 'Choosing Leisure: Social Theory, Class and Generations', in P. Bramham and S. Wagg (eds) *The New Politics of Leisure and Pleasure* (Basingstoke: Palgrave Macmillan).
Bramham, P. and Wagg, S. (2010) *The New Politics of Leisure and Pleasure* (Basingstoke: Palgrave Macmillan).
Braund, D. (1994) 'The Luxuries of Athenian Democracy', *Greece and Rome*, 41, 41–48.
Braund, S. (2004) *Juvenal and Perseus* (Cambridge, MA: Harvard, Loeb Classical Library).
Bray, W. and Dollery, C. (1983) 'Coca Chewing and High-Altitude Stress: A Spurious Correlation', *Current Anthropology*, 24, 269–75.
Bremer, F. (2009) *Puritanism* (Oxford: Oxford University Press).
Briggs, A. and Burke, P. (2009) *A Social History of The Media* (Cambridge: Polity).
Bringmann, K. (2007) *History of the Roman Republic* (Cambridge: Polity).
Brom, F. (1932) 'The Maya Ball-Game *pok-ta-pok* (called *tlachtli* by the Aztecs)', *Middle American Research Series Publications*, 4, 485–530.
Brook, T. and Blue, G. (2002) *China and Historical Capitalism: Genealogies of Sinological Knowledge* (Cambridge: Cambridge University Press).
Brown, J. (1985) 'The Evolution of Darwin's Theism', *Journal of the History of Biology*, 19, 1–45.
Brown, P. (1989) *The World of Late Antiquity* (London: Thames and Hudson).
Brown, S. (2008) *Cinema Anime* (Basingstoke: Palgrave Macmillan).
Brownell, S. (2009) *The 1904 Anthropology Days and Olympic Games: Sport, Race, and American Imperialism* (Lincoln: University of Nebraska Press).
Bryson, A. (1998) *From Courtesy to Civility: Changing Codes of Conduct in Early Modern England* (Oxford: Oxford University Press).
Burdsey, D. (2006) 'If I ever Play Football, Dad, can I Play for England or India: British Asians, Sport and Diasporic National Identities', *Sociology*, 40, 11–28.
Burke, P. (1999) *Popular Culture in Early Modern Europe* (Aldershot: Ashgate).
Butterfield, H. (1968[1931]) *The Whig Interpretation of History* (London: Bell).
Cameron, A. (1976) *Circus Factions* (Oxford: Clarendon).
Cameron, A. (1979) 'Images of Authority: **Elites and Icons** in Late Sixth-Century Byzantium', *Past and Present*, 84, 3–35.
Campbell, C. and Corns, T. (2010) *John Milton: Life, Work, and Thought* (Oxford: Oxford University Press).
Canfield, J. (2007) *Becoming Human: The Development of Language, Self and Self-consciousness between Hominid and Human* (Basingstoke: Palgrave Macmillan).
Carpenter, H. (2002) *J.R.R. Tolkien: A Biography* (London: Allen and Unwin).
Carpenter, H. (2006) *The Letters of J.R.R. Tolkien* (London: Allen and Unwin).
Carr, K. (1993) 'Making Way: War, Philosophy and Sport in Japanese Judo', *Journal of Sport History*, 20, 167–89.
Carrington, B. (2004) 'Cosmopolitan Olympism, Humanism and the Spectacle of "Race"' in J. Bale and M. Cristensen (eds) *Post-Olympism? Questioning Sport in the Twenty-First Century* (Oxford: Berg).
Carrington, B. and McDonald, I. (2001) *'Race', Sport and British Society* (London: Routledge).
Carrington, B. and McDonald, I. (2008) *Marxism, Cultural Studies and Sport* (London: Routledge).
Carruthers, M. (2010) *Rhetoric beyond Words: Delight and Persuasion in the Arts of the Middle Ages* (Cambridge: Cambridge University Press).

Cassius Dio (2005) *The Roman History: The Reign of Augustus*, translated by J. Carter and I. Scott-Kilvert (Harmondsworth: Penguin).
Cefalu, P. (2009) *Moral Identity in Early Modern English Literature* (Cambridge: Cambridge University Press).
Chadwick, H. (1993) *The Penguin History of the Church, vol.1: The Early Church* (Harmondsworth: Penguin).
Chaney, E. (2000) *The Evolution of the Grand Tour* (London: Routledge).
Chaucer, G. (2003) *The Canterbury Tales*, translated by N. Coghill (Harmondsworth: Penguin).
Chen, M. (2009) 'Communication, Coordination, and Camaraderie in World of Warcraft', *Games and Culture*, 4, 47–73.
Cicero (2002) *Selected Political Speeches*, translated by M. Grant (Harmondsworth: Penguin).
Clark, P. (2008) *The Chinese Cultural Revolution: A History* (Cambridge: Cambridge University Press).
Clarke, J. and Critcher, C. (1985) *The Devil Makes Work* (Basingstoke: Palgrave Macmillan).
Clements, J. (2004) *Confucius: A Biography* (Stroud: The History Press).
Coakley, J. (2003) *Sports in Society: Issues and Controversies* (New York: McGraw Hill).
Coalter, F. (1998) 'Leisure Studies, Leisure Policy and Social Citizenship: The Failure of Welfare or the Limits of Welfare?', *Leisure Studies*, 17, 21–36.
Coffey, J. and Lim, P. (2008) *The Cambridge Companion to Puritanism* (Cambridge: Cambridge University Press).
Cohodas, M. (1975) 'The Symbolism and Ritual Function of the Middle Classic Ball Game in Mesoamerica', *American Indian Quarterly*, 2, 99–130.
Cohen, D. (2006) *The Development of Play* (London: Routledge).
Cohen, H. (1994) *The Scientific Revolution: A Historiographical Inquiry* (Chicago: University of Chicago Press).
Cohen, P. (2003) *China Unbound: Evolving Perspectives on the Chinese Past* (London: Routledge).
Collins, H. and Pinch, T. (1993) *The Golem: What Everyone Should Know about Science* (Cambridge: Cambridge University Press).
Collins, T. (1999) *Rugby's Greatest Split* (London: Frank Cass).
Collins, T. (2006) *Rugby League in Twentieth Century Britain* (London: Routledge).
Collins, T and Vamplew, W. (2002) *Mud, Sweat and Beers: A Cultural History of Sport and Alcohol* (Oxford: Berg).
Comnena, A. (2003) *Alexiad*, translated by E. Sewter (Harmondsworth: Penguin).
Conkey, M. (1989) 'New Approaches in the Search for Meaning? A Review of Research in "Paleolithic Art"', *Journal of Field Archaeology*, 14, 413–30.
Connell, J. and Gibson, C. (2004) 'World Music: Deterritorializing Place and Identity', *Progress in Human Geography*, 28, 342–61.
Connell, R. (1995) *Masculinities* (Cambridge: Polity).
Cook, R. (2006) *Njal's Saga* (Harmondsworth: Penguin).
Copernicus, (1947[1543]) *De Revolutionibus Orbium Caelestium*, translated by Dobson and Brodetsky (London: Royal Astronomical Society).
Cord, S. (1985) *Henry George: Dreamer or Realist?* (New York: Schalkenbach).
Corn, A. (2010) 'Land, Song, Constitution: Exploring Expressions of Ancestral Agency, Intercultural Diplomacy and Family Legacy in the Music of Yothu Yindi with Mandawuy Yunupinu', *Popular Music*, 29, 81–102.

Courbin, P. (1988) *What Is Archaeology?* (Chicago: University of Chicago Press).

Cowan, E. (1989) 'The People of the Sunset: Some Recent Publications on Neolithic Britain', *Scottish Tradition*, 15, 1–23.

Crabbe, T. and Wagg, S. (2000) 'A Carnival of Cricket? The Cricket World Cup, "Race" and the Politics of Carnival', *Sport in Society*, 3, 70–88.

Crouch, D. (2006) *Tournament: A Chivalric Way of Life* (Winchester: Hambledon Continuum).

Crouch, D. (2010) *Flirting with Space: Journeys and Creativity* (Farnham: Ashgate).

Crowther, N. (1996) 'Sports Violence in the Roman and Byzantine Empires: A Modern Legacy?', *International Journal of the History of Sport*, 13, 445–58.

Cullen, L. (2003) *A History of Japan, 1582–1941: Internal and External Worlds* (Cambridge: Cambridge University Press).

Cunningham, A. and Jardine, N. (1990) *Romanticism and the Sciences* (Cambridge: Cambridge University Press).

D'Ambra, E. (2006) *Roman Women* (Cambridge: Cambridge University Press).

Danielson, D. (1999) *The Cambridge Companion to Milton* (Cambridge: Cambridge University Press).

Danziger, D. and Purcell, N. (2006) *Hadrian's Empire* (London: Hodder).

Darwin, C. (2009) *On the Origin of Species* (Harmondsworth: Penguin).

Darwin, J. (2009) *The Empire Project: The Rise and Fall of the British World-System, 1830–1970* (Cambridge: Cambridge University Press).

Debus, A. (1974) 'The Chemical Philosophers: Chemical Medicine from Paracelsus to Van Helmont', *History of Science*, 12, 235–59.

Debus, A. (1978) *Man and Nature in the Renaissance* (Cambridge: Cambridge University Press).

Deem, R. (1986) *All Work and No Play? The Sociology of Women and Leisure* (Milton Keynes: Open University Press).

Deem, R. (1999) 'How Do We Get Out of the Ghetto? Strategies for Research on Gender and Leisure for the Twenty-First Century', *Leisure Studies*, 18, 161–78.

Den Breejen, L. (2007) 'The Experiences of Long Distance Walking: A Case Study of the West Highland Way in Scotland', *Tourism Management*, 28, 1417–27.

Denham, D. (2004) 'Global and Local Influences on English Rugby League', *Sociology of Sport Journal*, 21, 206–19.

Depew, D. and Weber, B. (1995) Darwinism Evolving: Systems Dynamics and the Genealogy of Natural Selection (Cambridge: MIT Press).

Derrida, J. (1982) *Margins of Philosophy* (Chicago: University of Chicago Press).

Descartes, R. (2003) *Meditations and Other Metaphysical Writings* (Harmondsworth: Penguin).

Desmond, A. (1989) *The Politics of Evolution* (Chicago: University of Chicago Press).

Desmond, A. and Moore, J. (1991) *Darwin* (London: Michael Joseph).

Dettmar, K. (1997) *Reading Rock and Roll* (New York: Columbia University Press).

Dewey, J. (1916) *Democracy and Education* (New York: Free Press).

Dewey, J. (1933) *How We Think* (Boston: Houghton Miffin).

Diggins, J. (1999) *Thorstein Veblen: Theorist of the Leisure Class* (Princeton: Princeton University Press).

Diller, A. (1962) 'Photius' *Bibliotheca* in Byzantine Literature', *Dumbarton Oaks Papers*, 16, 389–96.

Dobbs, B. (1982) 'Newton's Alchemy and His Theory of Matter', *Isis*, 73, 511–28.

Donner, F. (2010) *Muhammad and the Believers: At the Origins of Islam* (Cambridge: Harvard University Press).

Douglas, M. (1991) *Purity and Danger: An Analysis of the Concepts of Pollution and Taboo* (London: Routledge).

Douglass, C. (1999) *Bulls, Bullfighting, and Spanish Identities* (Tucson: University of Arizona Press).

Drake, P. and Haynes, R. (2010) 'Television, Deregulation and the Reshaping of Leisure', in P. Bramham and S. Wagg (eds) *The New Politics of Leisure and Pleasure* (Basingstoke: Palgrave Macmillan).

Du Boulay, F. (1991) *The England of Piers Plowman: William Langland and His Vision of the Fourteenth Century* (Cambridge: Brewer).

Dudrah, R. (2006) *Bollywood: Sociology Goes to the Movies* (London: Sage).

Duhem, P. (1969) *To Save the Phenomena: An Essay on the Idea of Physical Theory from Plato to Galileo* (Chicago: University of Chicago Press).

Dunkle, R. (2008) *Gladiators: Violence and Spectacle in Ancient Rome* (London: Longman).

Dunning, E. and Sheard, K. (1979) *Barbarians, Gentlemen and Players* (Oxford: Martin Robertson).

Durha, C. and Pruitt, K. (2008) *Uncircumscribed Mind: Reading Milton Deeply* (Selinsgrove: Susquehanna University Press).

Durkheim, E. and Mauss, M. (1969) *Primitive Classification* (Chicago: University of Chicago Press).

Dyer, C. (1989) *Standards of Living in the Later Middle Ages: Social Change in England c.1200–1520* (Cambridge: Cambridge University Press).

Dyer, C. (1994) 'The English Medieval Village Community and Its Decline', *Journal of British Studies*, 33, 407–29.

Easthope, A. (1999) *Englishness and National Culture* (London: Routledge).

Echard, S. (2011) *The Arthur of Medieval Latin Literature: The Development and Dissemination of the Arthurian Legend in Medieval Latin* (Cardiff: University of Wales Press).

Eco, U. (1986) *Faith in Fakes* (London: Secker and Warburg).

Edensor, T. (2002) *National Identity: Popular Culture and Everyday Life* (Oxford: Berg).

Effros, B. (2002) *Creating Community with Food and Drink in Merovingian Gaul* (Basingstoke: Palgrave Macmillan).

Ehland, C. (2007) *Thinking Northern: Textures of Identity in the North of England* (Amsterdam: Rodopi).

Eisenstein, E. (1983) *The Printing Revolution in Early Modern Europe* (Cambridge: Cambridge University Press).

Elias, N. (1978) *The Civilizing Process: Volume One* (Oxford: Blackwell).

Elias, N. (1982) *The Civilizing Process: Volume Two* (Oxford: Blackwell).

Elias, N. and Dunning, E. (1986) *The Quest for Excitement* (Oxford: Blackwell).

Elsner, J. (1992) 'Pausanias: A Greek Pilgrim in a Roman World', *Past and Present*, 135, 3–29.

Evans, J. (2000) *The Age of Justinian: The Circumstances of Imperial Power* (London: Routledge).

Evans, R. (1997) *In Defence of History* (London: Granta).

Fagan, G. (2011) *The Lure of the Arena: Social Psychology and the Crowd at the Roman Games* (Cambridge: Cambridge University Press).

Fairbank, J. and Goldman, M. (2006) *China: A New History* (Cambridge: Harvard University Press).
Farooq, S. and Parker, A. (2009) 'Sport, Physical Education, and Islam: Muslim Independent Schooling and the Social Construction of Masculinities', *Sociology of Sport Journal*, 26, 277–95.
Faroqhi, S. (2005) *Subjects of the Sultan: Culture and Daily Life in the Ottoman Empire* (London: I.B. Taurus).
Fawbert, J. (2005) 'Football Fandom, West Ham United and the "Cockney Diaspora": From Working-Class Community to Youth Post-tribe?', in P. Bramham and J. Caudwell (eds) *Sport, Active Leisure and Youth Cultures* (Eastbourne: Leisure Studies Association).
Featherstone, M. (1990) *Global Culture: Nationalism, Globalization and Modernity* (London: Sage).
Fenlon, I. (2007) *The Ceremonial City: History, Memory and Myth in Renaissance Venice* (London: Yale University Press).
Fichman, M, 1997, 'Biology and Politics: Defining the Boundaries', in R. Lightman (ed.) *Victorian Science in Context* (London: University of Chicago Press).
Fimi, D. (2009) *Tolkien, Race and Cultural History* (Basingstoke: Palgrave Macmillan).
Fletcher, R. (2001) *Moorish Spain* (London: Phoenix).
Fletcher, R. (2004) *The Cross and the Crescent* (Harmondsworth: Penguin).
Foucault, M. (1970) *The Order of Things* (London: Tavistock).
Foucault, M. (1972) *The Archaeology of Knowledge* (London: Tavistock).
Fox, J. (1995) 'Ballcourts and Political Ritual in Southern Mesoamerica', *Current Anthropology*, 37, 483–96.
Fox, R.L. (2005) *The Classical World* (Harmondsworth: Penguin).
Fraleigh, W.P. (1984) *Right Actions in Sport: Ethics for Contestants* (Champaign: Human Kinetics Publishers).
Francmanis, J. (2002) 'National Music to National Redeemer: The Consolidation of a "Folk-Song" Construct in Edwardian England', *Popular Music*, 21, 1–25.
Frank, J., Moore, R. and Ames, G. (2000) 'Historical and Cultural Roots of Drinking Problems among American Indians', *American Journal of Public Health*, 90, 344–51.
Frazer, J. (2004[1890]) *The Golden Bough: A Study in Magic and Religion* (Cambridge: Cambridge University Press).
Freeman, C. (2009) *A New History of Early Christianity* (London: Yale University Press).
Friell, G. and Williams, S. (1998) *Theodosius: The Empire at Bay* (London: Routledge).
Fuller, S. (1992) 'Being There with Thomas Kuhn: A Parable for Postmodern Times', *History and Theory*, 31, 241–75.
Fung, E. (2000) *In Search of Chinese Democracy: Civil Opposition in Nationalist China, 1929–1949* (Cambridge: Cambridge University Press).
Fyfe, F. (1943) *The Art of Falconry* (Palo Alto: Stanford University Press).
Gamble, C. (2007) *Origins and Revolutions: Human Identity in Earliest Prehistory* (Cambridge: Cambridge University Press).
Gardner, M. (1999) *Kant and the Critique of Pure Reason* (London: Routledge).
Garry, J. (1983) 'The Literary History of the English Morris Dance', *Folklore*, 94, 219–28.
Geertz, C. (1973) *The Interpretation of Cultures* (New York: Basic).
Gellner, E. (2006) *Nations and Nationalism* (Oxford: Blackwell).

Geoffrey of Monmouth (2004) *The History of the Kings of Britain*, translated by L. Thorpe (Harmondsworth: Penguin).
George, H. (1979[1979]) *Progress and Poverty* (London: Hogarth Press).
Gibbon, E. (2005[1776–1788]) *The History of the Decline and Fall of the Roman Empire*, six volumes (reprinted in three volumes) (Harmondsworth: Penguin).
Gibson, J. (1993) *Performance versus Results* (Albany: State University of New York Press).
Giddens, A. (1990) *The Consequences of Modernity* (Cambridge: Polity).
Gimbutas, M. (1963) 'European Prehistory: Neolithic to the Iron Age', *Biennial Review of Anthropology*, 3, 69–106.
Ginzburg, C. (1992) *The Cheese and the Worms: The Cosmos of a Sixteenth-Century Miller* (Baltimore, MD: The Johns Hopkins University Press).
Giulianotti, R. and Robertson, R. (2007) 'Forms of Glocalization: Globalization and the Migration Strategies of Scottish Football Fans in North America', *Sociology*, 41, 133–52.
Glasier, P. (1998) *Falconry and Hawking* (London: Batsford).
Golden, M. (1998) *Sport and Society in Ancient Greece* (Cambridge: Cambridge University Press).
Goldhill, S. (2001) *Being Greek Under Rome* (Cambridge: Cambridge University Press).
Golinski, J. (1998) *Making Natural Knowledge: Constructivism and the History of Science* (Cambridge: Cambridge University Press).
Gooday, G. (2004) *The Morals of Measurement: Accuracy, Irony and Trust in Late Victorian Electrical Practice* (Cambridge: Cambridge University Press).
Goodwin, J. (2008) *Lords of the Horizons: A History of the Ottoman Empire* (London: Vintage).
Gootenberg, P. (1999) *Cocaine: Global Histories* (London: Routledge).
Gordon, A. (1998) *The Wages of Affluence: Labor and Management in Postwar Japan* (Cambridge: Harvard University Press).
Gosden, C. and Hather, J. (1999) *The Prehistory of Food* (London: Routledge).
Gossman, L. (2008) *The Empire Unpossess'd: An Essay on Gibbon's Decline and Fall* (Cambridge: Cambridge University Press).
Goto-Jones, C. (2009) *Modern Japan* (Oxford: Oxford University Press).
Gould, S.J. (1997) *Mismeasure of Man* (Harmondsworth: Penguin).
Grant, E. (1997) *The Foundations of Modern Science in the Middle Ages* (London: Routledge).
Gray, W. (2010) *Fantasy, Myth and the Measure of Truth* (Basingstoke: Palgrave Macmillan).
Greenaway, J. (2003) *Drink and British Politics since 1830* (Basingstoke: Palgrave Macmillan).
Gregory, T. (2005) *A History of Byzantium: 306–1453* (Oxford: Blackwell).
Griffin, M. (1987) *Nero: The End of a Dynasty* (Harmondsworth: Penguin).
Guralnick, P. (1995) *Last Train to Memphis* (London: Abacus).
Gurr, A. (2004) *Playgoing in Shakespeare's London* (Cambridge: Cambridge University Press).
Guttmann, A. (1981) 'Sports Spectators from Antiquity to the Renaissance', *Journal of Sports History*, 8, 5–27.
Guttmann, A. (1986) *Sports Spectators* (New York: Columbia University Press).

Guttmann, A. (1992) 'Chariot Races, Tournaments and the Civilizing Process', in E. Dunning and C. Rojek (eds) *Sport and Leisure in the Civilizing Process* (London: Palgrave Macmillan).
Guyer, P. (1992) *The Cambridge Companion to Kant* (Cambridge: Cambridge University Press).
Habermas, J. (1984[1981]) *The Theory of Communicative Action, Volume One: Reason and the Rationalization of Society* (Cambridge: Polity).
Habermas, J. (1987[1981]) *The Theory of Communicative Action, Volume Two: The Critique of Functionalist Reason* (Cambridge: Polity).
Habermas, J. (1989[1962]) *The Structural Transformation of the Public Sphere* (Cambridge: Polity).
Habermas, J. (1990) *The Philosophical Discourse of Modernity* (Cambridge: Polity).
Habermas, J. (1992) *Post-Metaphysical Thinking: Philosophical Essays* (Cambridge: Polity).
Haines, J. (2004) *Eight Centuries of Troubadours and Trouvères: The Changing Identity of Medieval Music* (Cambridge: Cambridge University Press).
Haldon, J. (2008) *A Social History of Byzantium* (Oxford: Blackwell).
Hall, S. (1993) 'Culture, Community, Nation', *Cultural Studies*, 7, 349–63.
Halsall, G. (2007) *Barbarian Migrations and the Roman West, 376–568* (Cambridge: Cambridge University Press).
Hammond, P. (2005) *Food and Feast in Medieval England* (Stroud: The History Press).
Hanioglu, S. (2008) *A Brief History of the Late Ottoman Empire* (Princeton: Princeton University Press).
Hankins, T. (1985) *Science and the Enlightenment* (Cambridge: Cambridge University Press).
Hannam, J. (2010) *God's Philosophers: How the Medieval World Laid the Foundations of Modern Science* (London: Icon).
Hargreaves, J. (1994) *Sporting Females: Critical Issues in the History and Sociology of Women's Sport* (London: Routledge).
Hark, I.R. (2008) *Star Trek* (Basingstoke: Palgrave Macmillan).
Harootunian, H. (2001) *Overcome by Modernity: History, Culture, and Community in Interwar Japan* (Princeton: Princeton University Press).
Harris, J. (2010) *The End of Byzantium* (London: Yale University Press).
Harris, R. (1990) *Language, Saussure and Wittgenstein: How to Play Games with Words* (London: Routledge).
Hart, J. (2008) *Comparing Empires* (Basingstoke: Palgrave Macmillan).
Harvey, D. (1989) *The Condition of Postmodernity* (Oxford: Blackwell).
Hayman, R. (1980) *Nietzsche: A Critical Life* (Harmondsworth: Penguin).
Hazelwood, N. (2001) *Savage: Survival, Revenge and the Theory of Evolution* (London: Hodder and Stoughton).
Heather, P. (1998) *The Goths* (Oxford: Blackwell).
Heather, P. (2006) *The Fall of the Roman Empire: A New History* (London: Pan).
Hellman, R. (1987) *Henry George Reconsidered* (New York: Carlton).
Hendon, J. and Joyce, R. (2003) *Mesoamerican Archaeology* (New York: Wiley-Blackwell).
Hendry, J. (2003) *Understanding Japanese Society* (London: Routledge).
Henry, J. (1990) 'Magic and Science in the Sixteenth and Seventeenth Centuries', in R. Olby, G. Cantor, J. Christie and J. Hodge (eds) *Companion to the History of Modern Science* (London: Routledge).

Hill, C. (1991) *The World Turned Upside Down: Radical Ideas during the English Revolution* (Harmondsworth: Penguin).
Hjorth, L. (2001) *Games and Gaming* (Oxford: Berg).
Hobbes, T. (2002) *Leviathan* (Harmondsworth: Penguin).
Hobsbawm, E. (1988) *The Age of Capital* (London: Abacus).
Hobsbawm, E. (1989) *The Age of Empire* (London: Abacus).
Hobsbawm, E. (1992) *Nations and Nationalism since 1780* (Cambridge: Cambridge University Press).
Hobsbawm, E. (1995) *Age of Extremes* (London: Abacus).
Hobsbawm, E. and Ranger, T. (1983) *The Invention of Tradition* (Cambridge: Cambridge University Press).
Hodder, I. (1995) *Theory and Practice in Archaeology* (London: Routledge).
Hofstadter, R. (1955) *Social Darwinism in American Thought 1860–1915* (Boston, Beacon).
Hoggart, R. (1957) *The Uses of Literacy* (London: Chatto and Windus).
Holowchak, M. (2007) 'Games As Pastimes in Suits's Utopia: Meaningful Living and the "Metaphysics of Leisure"', *Journal of the Philosophy of Sport*, 34, 88–96.
Holst, G. (1977) *Road to Rembetika: Music from a Greek Sub-Culture* (Athens: Denise Harvey).
Holt, R. (1989) *Sport and the British: A Modern History* (Oxford: Clarendon).
Holton, R. (2008) *Global Networks* (Basingstoke: Palgrave Macmillan).
Horne, J. (2006) *Sport in Consumer Culture* (Basingstoke: Palgrave Macmillan).
Horodowich, E. (2005) 'The Gossiping Tongue: Oral Networks, Public Life and Political Culture in Early Modern Venice', *Renaissance Studies*, 19, 22–45.
Hourani, A. (2005) *A History of the Arab Peoples* (London: Faber and Faber).
Houston, S. (2008) *The First Writing: Script Invention As History and Process* (Cambridge: Cambridge University Press).
Howard, J. (1988) 'Crossdressing, the Theatre, and Gender Struggle in Early Modern England', *Shakespeare Quarterly*, 39, 418–40.
Hudson, R. (2003) 'Novelty and the 1919 Eclipse Experiments', *Studies in History and Philosophy of Science Part B: Studies in History and Philosophy of Modern Physics*, 34, 107–29.
Hughes, A. (1998) *The Causes of the English Civil War* (Basingstoke: Palgrave Macmillan).
Hughes, B. (2001) *Evolutionary Playwork and Reflective Analytic Practice* (London: Routledge).
Huizinga, J. (2003[1944]) *Homo Ludens: A Study of the Play-Element in Culture* (London: Taylor and Francis).
Humphrey, N. (1998) 'Cave Art, Autism, and the Evolution of the Human Mind', *Cambridge Archaeological Journal*, 8, 165–91.
Hunter, J. (1989) *The Emergence of Modern Japan: An Introductory History since 1853* (London: Longman).
Huppert, G. (1998) *After the Black Death: A Social History of Early Modern Europe* (Bloomington: Indiana University Press).
Hussey, J. (1935) 'Michael Psellus: The Byzantine Historian', *Speculum*, 10–81-90.
Hutchings, G. (2000) *Modern China: Companion to a Rising Power* (Harmondsworth: Penguin).
Hutton, R (2006) *Witches, Druids and King Arthur* (Winchester: Hambledon Continuum).

Huxley, A. (2007[1932]) *Brave New World* (London: Vintage).
Huxley, T.H. (2001) *Man's Place in Nature* (New York: Modern Library).
Hylton, K. (2005) '"Race", Sport and Leisure: Lessons from Critical Race Theory', *Leisure Studies*, 24, 81–98.
Inalcik, H. (2000) *The Ottoman Empire 1300–1600* (London: Phoenix).
Irvine, W. (2008) *A Guide to the Good Life: The Ancient Art of Stoic Joy* (Oxford: Oxford University Press).
Iwabuchi, K. (2003) *Recentering Globalization: Popular Culture and Japanese Transnationalism* (Durham: Duke University Press).
Jacobson, J. (1997) 'Religion and Ethnicity: Dual and Alternative Sources of Identity among Young British Pakistanis', *Ethnic and Racial Studies*, 20, 238–56.
James, C.L.R. (2005[1963]) *Beyond a Boundary* (London: Yellow Jersey Press).
Jansen, M. (2002) *The Making of Modern Japan* (Cambridge: Harvard University Press).
Jardine, L. (2009) *Going Dutch* (London: Harper Perennial).
Jenkins, K. (1991) *Re-thinking History* (London: Routledge).
Jenkins, K. (1995) *On 'What Is History?' – From Carr and Elton to Rorty and White* (London: Routledge).
Jenkins, K. (1997) *The Postmodern History Reader* (London: Routledge).
Jochens, J. (1998) *Women in Old Norse Society* (Ithaca, New York: Cornell University Press).
Jones, G. (1984) *A History of the Vikings* (Oxford: Oxford University Press).
Jones, H. (1979) *The Epicurean Tradition* (London: Routledge).
Jones, L. (1945) 'The Influence of Cassiodorus on Mediaeval Culture', *Speculum*, 20, 433–22.
Jordanova, L. (2000) *History in Practice* (London: Arnold).
Joyce, P. (1995) 'The End of Social History?', *Social History*, 20, 73–91.
Kaegi, W. (2003) *Heraclius, Emperor of Byzantium* (Cambridge: Cambridge University Press).
Kant, I. (2007[1781]) *Critique of Pure Reason*, translated by M. Muller (Harmondsworth: Penguin).
Kant, I. (2007[1790]) *Critique of Judgement*, translated by J.C. Meredith (Oxford: Oxford University Press).
Karabell, Z. (2007) *People of the Book: The Forgotten History of Islam and the West* (London: John Murray).
Kastan, D. (1986) 'Proud Majesty Made a Subject: Shakespeare and the Spectacle of Rule', *Shakespeare Quarterly*, 37, 459–75.
Kater, M. (2006) *Hitler Youth* (Cambridge: Harvard University Press).
Kavoori, A. and Punathambekar, A. (2008) *Global Bollywood* (New York: New York University Press).
Keene, D. (2006) *Yoshimasa and the Silver Pavilion: The Creation of the Soul of Japan* (New York: Columbia University Press).
Kelner, S. (1996) *To Jerusalem and Back* (Basingstoke: Palgrave Macmillan).
Kemp, B. (2005) *Ancient Egypt* (London: Routledge).
Kennedy, H. (1996) *Muslim Spain and Portugal: Political History of Al-Andalus* (London: Longman).
Kennedy, H. (2004) *The Prophet and the Age of the Caliphates: The Islamic Near East from the 6th to the 11th Century* (London: Longman).

Kidd, T. (2005) *The Protestant Interest: New England after Puritanism* (London: Yale University Press).
Kieckhefer, R. (1989) *Magic in the Middle Ages* (Cambridge: Cambridge University Press).
King, C. (2007) 'Staging the Winter *White Olympics* or, Why Sport Matters to *White* Power', *Journal of Sport and Social Issues*, 31, 89–94.
Ko, D. (2003) *Women and Confucian Cultures in Premodern China, Korea, and Japan* (Berkeley: University of California Press).
Kochin, M. (2002) *Gender and Rhetoric in Plato's Political Thought* (Cambridge: Cambridge University Press).
Kohn, M. (1995) *The Race Gallery* (London: Verso).
Kohl, P. (2009) *The Making of Bronze Age Eurasia* (Cambridge: Cambridge University Press).
Kuhn, T. (1957) *The Copernican Revolution* (Chicago: University of Chicago Press).
Kuusela, O. (2008) *Struggle against Dogmatism: Wittgenstein and the Concept of Philosophy* (Cambridge: Harvard University Press).
Kymlicka, W. (2002) *Contemporary Political Philosophy* (Oxford: Oxford University Press).
Langland, W. (1978) *The Vision of Piers Plowman* (London: Dent).
Lapidus, I. (2002) *A History of Islamic Societies* (Cambridge: Cambridge University Press).
Latour, B. (1987) *Science in Action* (Cambridge: Harvard University Press).
Lee, S. (2005) *Russia and the USSR, 1855–1991: Autocracy and Dictatorship* (London: Routledge).
Lees, C. (1994) *Medieval Masculinities: Regarding Men in the Middle Ages* (Minneapolis: University of Minnesota Press).
Leiter, B. (2002) *Nietzsche on Morality* (London: Routledge).
Leroi-Gourhan, A. (1968) *The Art of Prehistoric Man in Ancient Europe* (London: Thames and Hudson).
Levi-Strauss, C. (1963) *Structural Anthropology* (New York: Basic).
Lewis-Williams, J. (2004) *The Mind in the Cave: Consciousness and the Origins of Art* (London: Thames and Hudson).
Lewis-Williams, J. and Dowson, T. (1988) 'The Signs of All Times: Entoptic Phenomena in Upper Palaeolithic Art', *Current Anthropology*, 2, 201–17.
Lightman, R. (1997) *Victorian Science in Context* (Chicago: University of Chicago Press).
Lindberg, C. (2009) *The European Reformations* (Oxford: Blackwell).
Linnane, F. (2007) *London: The Wicked City* (London: Robson).
Lissner, D. and Lissner, W. (1992) *George and Democracy in the British Isles* (New York: Schalkenbach).
Liudprand (2007) *The Complete Works*, translated by P. Squatriti (Washington: Catholic University of America Press).
Livy (2005) *The Early History of Rome*, translated by R. Ogilvie and A. De Selincourt (Harmondsworth: Penguin).
Lloyd, G. and Sivin, N. (2004) *The Way and the Word: Science and Medicine in Early China and Greece* (London: Yale University Press).
Lorenz, C. (1994) 'Historical Knowledge and Historical Reality: A Plea for Internal Realism', *History and Theory*, 33, 297–327.

Lowney, C. (2006) *A Vanished World: Muslims, Christians, and Jews in Medieval Spain* (Oxford: Oxford University Press).
Loy, J.W. (1968) 'The Nature of Sport: A Definitional Effort', *QUEST*, 10, 1–15.
Maalouf, A. (1984) *The Crusades through Arab Eyes* (London: Saqi Books).
MacCannell, D. (1973) 'Staged Authenticity: Arrangements of Social Space in Tourist Settings', *American Journal of Sociology*, 79, 589–603.
MacCannell, D. (1976) *The Tourist: A New Theory of the Leisure Class* (New York: Schoken Books).
Mackay, C. (2003[1841]) *Extraordinary Popular Delusions and the Madness of Crowds* (Petersfield: Harriman House).
MacLean, C. (2010) 'Focus on India', in I. Ronde (ed.) *Malt Whisky Yearbook 2011* (Shrewsbury: MagDig Media).
MacNamee, M. (2007) 'Sport, Ethics and Philosophy: Contexts, History, Prospects', *Sport, Ethics and Philosophy*, 1, 1–6.
MacNeilage, P. (2010) *The Origin of Speech* (Oxford: Oxford University Press).
Macrory, J. (1991) *Running with the Ball* (London: Collins Willow).
Maczak, A. (1995) *Travel in Early Modern Europe* (New York: Wiley-Blackwell).
Maguire, J. (2005) *Power and Global Sport* (London: Routledge).
Malory, T. (2008[1470]) *Le Morte Darthur* (Oxford: Oxford University Press).
Mangan, J. (1981) *Athleticism in the Victorian and Edwardian Public Schools* (Cambridge: Cambridge University Press).
Mangan, J. (1995) 'Duty unto Death: English Masculinity and Militarism in the age of the New Imperialism', in J. Mangan (ed.) *Tribal Identities: Nationalism, Europe, Sport* (London: Frank Cass).
Mangan, J. and Ritchie, A. (2005) *Ethnicity, Sport, Identity: Struggles for Status* (London: Routledge).
Mango, C. (1980) *Byzantium* (London: Weidenfeld and Nicolson).
Mango, C. (2000) *Byzantium* (London: Pimlico).
Mann, J. (1973) *Chaucer and Medieval Estates Satire: The Literature of Social Classes and the General Prologue to the Canterbury Tales* (Cambridge: Cambridge University Press).
Manuel, P. (1993) *Cassette Culture: Popular Music and Technology in North India* (Chicago: University of Chicago Press).
Marchant, J. (1975) *Alfred Russell Wallace: Letters and Reminisces* (New York: Arno Press).
Marcus Aurelius Antoninus (2006) *Meditations*, translated by M. Hammond (Harmondsworth: Penguin).
Marcus, G. and Fischer, M. (1986) *Anthropology As Cultural Critique* (Chicago: University of Chicago Press).
Marfany, J. (1997) 'The Invention of Leisure in Early Modern Europe', *Past and Present*, 156, 174–91.
Marqusee, M. (1994) *Anyone but England: Cricket and the National Malaise* (London: Verso).
Marqusee, M. (1999) 'In Search of the Unequivocal Englishman: The Conundrum of Race and Nation in English Cricket', in B. Carrington, and I. McDonald (eds) *'Race', Sport and British society* (London: Routledge).
Martin, C. (2010) *Milton among the Puritans* (Farnham: Ashgate).
Martinez, D. (1998) *The Worlds of Japanese Popular Culture: Gender, Shifting Boundaries and Global Cultures* (Cambridge: Cambridge University Press).

Maruyama, N., Yen, T-H. and Stronza, A. (2008) 'Perception of Authenticity of Tourist Art among Native American Artists in Santa Fe, New Mexico', *International Journal of Tourism Research*, 10, 453–66.
Marx, K. (1992[1867]) *Capital* (Harmondsworth: Penguin).
Marx, K. and Engels, F. (2004[1848]) *The Communist Manifesto* (Harmondsworth: Penguin).
Maslow, A. (1998) *Toward a Psychology of Being* (New York: John Wiley).
Masschaele, J. (2002) 'The Public Space of the Marketplace in Medieval England', *Speculum*, 77, 383–421.
Matarasso, P. (1975) *Quest of the Holy Grail* (Harmondsworth: Penguin).
Matheson, C. (2008) 'Music, Emotion and Authenticity: A Study of Celtic Music Festival Consumers', *Journal of Tourism and Cultural Change*, 6, 57–74.
Mattingly, D. (2007) *An Imperial Possession: Britain in the Roman Empire* (Harmondsworth: Penguin).
Mauss, M. (1990) *The Gift* (London: Routledge).
Maxwell, N. (1972) 'A Critique of Popper's Views on Scientific Method', *Philosophy of Science*, 39, 131–52.
Mazzotta, G. (1993) *The Worlds of Petrarch* (Durham: Duke University Press).
McBride, F. (1988) 'A Critique of Mr. Suits' Definition of Game Playing', in W. Morgan and K. Meier (eds) *Philosophic Inquiry in Sport* (Champaign: Human Kinetics).
McDonald, J.I.H. (1998) *The Crucible of Christian Morality* (London: Routledge).
McDonnell, M. (2009) *Roman Manliness: 'Virtus' and the Roman Republic* (Cambridge: Cambridge University Press).
McFee, G. (2004) *Sport, Rules and Values* (London: Routledge).
McGovern, P. (2011) *Uncorking the Past: The Quest for Wine, Beer, and Other Alcoholic Beverages* (Berkeley: University of California Press).
McGrew, W. (2004) *The Cultured Chimpanzee: Reflections on Cultural Primatology* (Cambridge: Cambridge University Press).
McGuigan, J. (2006) *Modernity and Postmodern Culture* (Maidenhead: Open University Press).
McKitterick, R. (2008) *Charlemagne: The Formation of a European Identity* (Cambridge: Cambridge University Press).
Mead, M. (2001) *Coming of Age in Samoa* (London: Harper Perennial).
Mehta, N., Gemmell, J. and Malcolm, D. (2009) 'Bombay Sport Exchange: Cricket, Globalization and the Future', *Sport in Society*, 12, 694–707.
Melling, P. and Collins, T. (2004) *The Glory of Their Times* (Skipton: Vertical).
Mellor, R. (1998) *The Roman Historians* (London: Routledge).
Melton, J. (2001) *The Rise of the Public in Enlightenment Europe* (Cambridge: Cambridge University Press).
Mill, J.S. (1998[1859]) *On Liberty* (Oxford: Oxford University Press).
Milton, J. (2003[1667]) *Paradise Lost* (Harmondsworth: Penguin).
Mishra, V. (2002) *Bollywood Cinema: Temples of Desire* (London: Routledge).
Mitter, P. (2001) *Indian Art* (Oxford: Oxford University Press).
Mitter, R. (2004) *A Bitter Revolution: China's Struggle with the Modern World* (Oxford: Oxford University Press).
Mitter, R. (2008) *Modern China* (Oxford: Oxford University Press).
Monaghan, J. and Just, P. (2001) *Social and Cultural Anthropology: A Very Short Introduction* (Oxford: Oxford University Press).

Moore, R. (2000) *The First European Revolution: 970–1215* (Oxford: Blackwell).
Moorhouse, G. (1989) *At the George* (Sevenoaks: Hodder and Stoughton).
Morgan, W. (1976) 'On the Path – Towards an Ontology of Sport', *Journal of the Philosophy of Sport*, 3, 25–34.
Morgan, W. (2005) *Why Sports Morally Matter* (London: Routledge).
Morgan, W. (2008) 'Some Further Words on Suits on Play', *Journal of the Philosophy of Sport*, 35, 120–41.
Morrow, S. (2005) 'Continuity and Change: The Planning and Management of Long Distance Walking Routes in Scotland', *Managing Leisure*, 10, 237–50.
Moylan, T. (2000) *Scraps of the Untainted: Science Fiction, Utopia, Dystopia* (New York: Perseus).
Muir, E. (1986) *Civic Ritual in Renaissance Venice* (Princeton: Princeton University Press).
Mukerji, C. (1994) 'The Political Mobilization of Nature in Seventeenth-Century French Formal Gardens', *Theory and Society*, 23, 651–77.
Murray, M. (1921) *Witchcraft in Modern Europe* (Oxford: Clarendon).
Napier, S. (2008) *From Impressionism to Anime: Japan As Fantasy and Fan Cult in the Mind of the West: Japan As Fantasy and Fan Culture in the Mind of the West* (Basingstoke: Palgrave Macmillan).
Norwich, J. (1982) *A History of Venice* (London: Vintage).
Obolensky, D. (2000) *Byzantine Commonwealth: Eastern Europe, 500–1453* (London: Weidenfeld and Nicolson).
O'Brien, K. (1997) *Narratives of Enlightenment: Cosmopolitan History from Voltaire to Gibbon* (Cambridge: Cambridge University Press).
Oggins, R. (2004) *The Kings and Their Hawks* (London: Yale University Press).
O'Keefe, T. (2009) *Epicureanism* (London: Acumen).
Oliver, P. (1990) *Blues Fell This Morning* (Cambridge: Cambridge University Press).
Orlin, L. (2009) *The Renaissance: A Sourcebook* (Basingstoke: Palgrave Macmillan).
Orwell, G. (2008[1949]) *Nineteen Eighty-Four* (Harmondsworth: Penguin).
Owens, J. (1981) 'Aristotle on Leisure', *Canadian Journal of Philosophy*, 11, 713–23.
Palmer, B. (2005) 'Early Modern Mobility: Players, Payments, and Patrons', *Shakespeare Quarterly*, 56, 259–305.
Palmer, C. and Brady, E. (2007) 'Landscape and Value in the Work of Alfred Wainwright (1907–1991)', *Landscape Research*, 32, 397–421.
Papakonstantinou, Z. (2002) 'Prizes in Early Archaic Greek Sport', *Nikephoros*, 15, 51–67.
Papakonstantinou, Z. (2003) 'Alcibiades in Olympia: Olympic Ideology, Sport and Social Conflict in Classical Athens', *Journal of Sport History*, 30, 173–82.
Papakonstantinou, Z. (2010) *Sport in the Cultures of the Ancient World: New Perspectives* (London: Routledge).
Parker, S. (1972) *The Future of Work and Leisure* (London: Paladin).
Parker, S. (1976) *The Sociology of Leisure* (London: Allen and Unwin).
Paterson, M. (1995) *The World of the Troubadours: Medieval Occitan Society, c.1100–c.1300* (Cambridge: Cambridge University Press).
Paul, C. (2008) *The Borghese Collections and the Display of Art in the Age of the Grand Tour* (Farnham: Ashgate).
Peake, M. (1946) *Titus Groan* (London: Eyre and Spottiswoode).
Peake, M. (1950) *Gormenghast* (London: Eyre and Spottiswoode).

Perloff, M. (1999) *Wittgenstein's Ladder: Poetic Language and the Strangeness of the Ordinary* (Chicago: University of Chicago Press).
Peters, F. (2007) *The Voice, the Word, the Books* (Princeton: Princeton University Press).
Petitt, T. (1984) 'Here Comes I, Jack Straw: English Folk Drama and Social Revolt', *Folklore*, 95, 3–20.
Petley, J. (2010) 'Doublethink: The Confused and Contradictory Politics of Television Censorship in Britain', in P. Bramham and S. Wagg (eds) *The New Politics of Leisure and Pleasure* (Basingstoke: Palgrave Macmillan).
Plato (2007) *The Republic*, translated by H.D.P. Lee and D. Lee (Harmondsworth: Penguin).
Pomeranz, K. (2000) *The Great Divergence: Chine, Europe and the Making of the Modern World Economy* (Princeton: Princeton University Press).
Popper, K. (1961) *The Logic of Scientific Discovery* (New York: Basic).
Porter, R. (1988) *Gibbon: Making History* (London: Phoenix).
Post, L. (1930) *The Prophet of San Francisco: Personal Memories and Interpretations of Henry George* (New York: Vanguard).
Pretzler, M. (2004) 'Turning Travel into Text: Pausanias at Work', *Greece and Rome*, 51, 199–216.
Pringle, R. and Markula, P. (2005) 'No Pain is Sane after All: A Foucauldian Analysis of Masculinities and Men's Experiences in Rugby', *Sociology of Sport*, 22, 475–97.
Procopius (1981) *The Secret History*, translated by G. Williamson (Harmondsworth: Penguin).
Pumfrey S., Slawinski, M. and Rossi, P. (1991) *Science, Culture and Popular Belief in Renaissance Europe* (Manchester: Manchester University Press).
Rawls, J. (1971) *A Theory of Justice* (New York: Routledge).
Read, P. (2003) *The Templars* (London: Phoenix).
Reader, I. (1989) 'Sumo: The Recent History of an Ethical Model for Japanese Society', *International Journal for the History of Sport*, 6, 285–98.
Reid, A. (2009) *Imperial Alchemy: Nationalism and Political Identity in Southeast Asia* (Cambridge: Cambridge University Press).
Renfrew, C. (2008) *Prehistory: The Making of the Human Mind* (London: Phoenix).
Restall, M. (2004) *Seven Myths of the Spanish Conquest* (Oxford: Oxford University Press).
Rhodes, P.J. (2003) 'Nothing to Do with Democracy: Greek Drama and the Polis', *The Journal of Hellenic Studies*, 123, 104–19.
Richardson, B. (2004) *Print Culture in Renaissance Italy: The Editor and the Vernacular Text, 1470–1600* (Cambridge: Cambridge University Press).
Riordan, J. and Jones, R. (1999) *Sport and Physical Education in China* (London: Taylor and Francis).
Roberts, K. (1978) *Contemporary Society and the Growth of Leisure* (London: Longman).
Roberts, K. (1999) *Leisure in Contemporary Society* (Wallingford: CAB International).
Roberts, K. (2000) 'The Impact of Leisure on Society', *World Leisure Journal*, 42, 3–10.
Roberts, K. (2004) *The Leisure Industries* (Basingstoke: Palgrave Macmillan).
Roberts, K. (2011) 'Leisure: The Importance of being Inconsequential', *Leisure Studies*, 30, 5–20.

Robinson, A. (2007) *The Story of Writing: Alphabets, Hieroglyphs and Pictograms* (London: Thames and Hudson).
Roche, M. (2006) 'Mega-events and Modernity Revisited: Globalization and the Case of the Olympics', *Sociological Review*, 54, 25–40.
Rodrigue, C. (1992) 'Can Religion Account for Early Animal Domestications?', *Professional Geographer*, 44, 417–30.
Rojek, C. (2002) 'Civil Labour, Leisure and Post Work Society', *Society and Leisure*, 25, 21–35.
Rojek, C. (2005) *Leisure Theory: Principles and Practice* (Basingstoke: Palgrave Macmillan).
Rojek, C. (2010) *The Labour of Leisure* (London: Sage).
Rojek, C. and Urry, J. (1997) *Touring Cultures: Transformations of Travel and Theory* (London: Routledge).
Rollefson, G. (1992) 'A Neolithic Game Board from Ain Ghazal, Jordan', *Bulletin of the American Schools of Oriental Research*, 286, 1–5.
Rorty, R. (1989) *Contingency, Irony and Solidarity* (Cambridge: Cambridge University Press).
Rose, P. (1975) *The Italian Renaissance of Mathematics: Studies on Humanists and Mathematicians from Petrarch to Galileo* (Geneva: Droz).
Rostworowski, M. (1998) *History of the Inca Realm* (Cambridge: Cambridge University Press).
Rousseau, J.J. (2002) *The Social Contract*, translated by S. Dunn (London: Yale University Press).
Rowe, W. (2001) *Saving the World: Chen Hongmou and Elite Consciousness in Eighteenth-Century China* (Palo Alto: Stanford University Press).
Runciman, S. (1992) *The Sicilian Vespers: A History of the Mediterranean World in the Later Thirteenth Century* (Cambridge: Cambridge University Press).
Ruse, M. (1979) *The Darwinian Revolution: Science Red in Tooth and Claw* (Chicago: University of Chicago Press).
Rutherford, J. (1997) *Forever England: Reflections on Masculinity and Empire* (London: Lawrence and Wishart).
Ruthven, M. (1997) *Islam: A Very Short Introduction* (Oxford: Oxford University Press).
Ryrie, A. (2009) *The Age of Reformation: The Tudor and Stewart Realms 1485–1603* (London: Longman).
Sachs, C. (1938) 'Towards a Prehistory of Occidental Music', *The Musical Quarterly*, 24, 147–52.
Saeki, T. (1994) 'The Conflict between Tradition and Modernization in a Sport Organization: A Sociological Study of Issues Surrounding the Organizational Reformation of the All Japan Judo Federation', *International Review for the Sociology of Sport*, 29, 301–15.
Sagona, A. and Zimansky, P. (2009) *Ancient Turkey* (London: Routledge).
Said, E. (1978) *Orientalism* (New York: Vintage).
Said, E. (1985) 'Orientalism Reconsidered', *Cultural Critique*, 1, 89–107.
Salamone, F. (2000) *Society, Culture, Leisure, and Play: An Anthropological Reference* (Lanham: University Press of America).
Scheiber, L. and Mitchell, M. (2010) Across a Great Divide: Continuity and Change in Native North American Societies, 1400–1900 (Tucson: University of Arizona Press).

Scribner, B. (1978) 'Reformation, Carnival and the World Turned Upside-Down', *Social History*, 3, 303–29.
Scullard, H.H. (1982) *From the Gracchi to Nero* (London: Routledge).
Searle, C. (2001) *Pitch of life: Writings on Cricket* (Manchester: Parrs Wood Press).
Shapin, S. (1998) *The Scientific Revolution* (Chicago: University of Chicago Press).
Shaw, D. (1978) 'Climate, Environment and Prehistory in the Sahara', *World Archaeology*, 8, 133–48.
Shaw, W. (1936) 'Rock Paintings in the Libyan Desert', *Antiquity*, 9, 175–78.
Shefrin, E. (2004) '*Lord of the Rings, Star Wars*, and Participatory Fandom: Mapping New Congruencies between the Internet and Media Entertainment Culture', *Critical Studies in Media Communication*, 21, 261–81.
Shepard, L., White, S. and Bruckner, M. (2000) *Songs of the Women Troubadours* (London: Routledge).
Shippey, T. (2005) *The Road to Middle-Earth* (London: HarperCollins).
Silverstein, J. (2010) *Islamic History: A Very Short Introduction* (Oxford: Oxford University Press).
Singh, K. (2001) *Sublimation* (London: Icon).
Siu, H. (1990) 'Recycling Tradition: Culture, History, and Political Economy in the Chrysanthemum Festivals of South China', *Comparative Studies in Society and History*, 32, 765–94.
Skinner, Q. (1969) 'Meaning and Understanding in the History of Ideas', *History and Theory*, 8, 3–53.
Skinner, R. (2010) 'Civil Taxis and Wild Trucks: The Dialectics of Social Space and Subjectivity in Dimanche à Bamako', *Popular Music*, 29, 17–39.
Smigel, E. (1963) *Work and Leisure: A Contemporary Social Problem* (New Haven, CT: College and University Press).
Smith, A. (1998) *Nationalism and Modernism* (London: Routledge).
Smith, M. (2002) *The Aztecs* (Oxford: Blackwell).
Smith, M. and Schreiber, K., (2006) 'New World States and Empires: Politics, Religion, and Urbanism', *Journal of Archaeological Research*, 14, 1–52.
Snape, R. (2004) 'The Co-operative Holidays Association and the Cultural Formation of Countryside Leisure Practice', *Leisure Studies*, 23, 143–58.
Snobelen, S. (1999) 'Isaac Newton, heretic: The Strategies of a Nicodemite', *British Journal for the History of Science*, 32, 381–419.
Snodgrass, A. (2003) 'Another Early Reader of Pausanias?', *The Journal of Hellenic Studies*, 123, 187–89.
Solomos, J. (1998) 'Beyond Racism and Multiculturalism', *Patterns of Prejudice*, 32, 45–62.
Solopova, E. and Lee, S. (2007) *Key Concepts in Medieval Literature* (Basingstoke: Palgrave Macmillan).
Soloway, R. (1982) 'Counting the Degenerates: The Statistics of Race Deterioration in Edwardian England', *Journal of Contemporary History*, 17, 137–64.
Southern, P. (2001) *Augustus* (London: Routledge).
Southern, R. (1990) *The Penguin History of the Church, vol. 1: Western Society and the Church in the Middle Ages* (Harmondsworth: Penguin).
Spracklen, K. (1996) *Playing the Ball: Constructing Community and Masculine Identity in Rugby* (unpublished PhD thesis: Leeds Metropolitan University).
Spracklen, K. (2006) 'Leisure, Consumption and a Blaze in the Northern Sky: Developing an Understanding of Leisure at the End of Modernity through the

Habermasian Framework of Communicative and Instrumental Rationality', *World Leisure Journal*, 48, 33–44.

Spracklen, K. (2007) 'Negotiations of Belonging: Habermasian Stories of Minority Ethnic Rugby League Players in London and the South of England', *World Leisure Journal*, 49, 216–26.

Spracklen, K. (2009) *The Meaning and Purpose of Leisure* (Basingstoke: Palgrave Macmillan).

Spracklen, K. and Spracklen, C. (2008) 'Negotiations of Being and Becoming: Minority Ethnic Rugby League Players in the Cathar Country of France', *International Review for the Sociology of Sport*, 43, 201–18.

Spracklen, K. and Fletcher, T. (2010) '"They'll Never Play Rugby League in Kazakhstan": Expansion, Community and Identity in a Globalised and Globalising Sport', in J. Caudwell (ed.) *Tourism and Leisure: Local Communities and Local Cultures* (Eastbourne: Leisure Studies Association).

Spracklen, K., Hylton, K. and Long, J. (2006) 'Managing and Monitoring Equality and Diversity in UK Sport: An Evaluation of the Sporting Equals Racial Equality Standard and its Impact on Organizational Change', *Journal of Sport and Social Issues*, 30, 289–305.

Spracklen, K., Timmins, S. and Long, J. (2010) 'Ethnographies of the Imagined, the Imaginary and the Critically Real: Blackness, Whiteness, the North of England and Rugby League', *Leisure Studies*, 29, 397–414.

Spurr, J. (1998) *English Puritanism* (Basingstoke: Palgrave Macmillan).

Stebbins, R. (1982) 'Serious Leisure: A Conceptual Statement', *Pacific Sociological Review*, 25, 251–72.

Stebbins, R. (2009) *Leisure and Consumption* (Basingstoke: Palgrave Macmillan).

Stenton, F. (1970) *Anglo-Saxon England* (Oxford: Oxford University Press).

Stephenson, T. (1935) 'Wanted – A Long Green Trail', *Daily Herald*, 22 June 1935, p. 17.

Stephenson, T. (1989) *Forbidden Land: The Struggle for Access to Mountain and Moorland* (Manchester: Manchester University Press).

Stern, P. and Stevenson, L. (2006) *Critical Inuit Studies* (Lincoln: University of Nebraska Press).

Stevens, A. (1975) 'Animals in Palaeolithic Cave Art: Leroi-Gourhan's Hypothesis', *Antiquity*, 49, 54–57.

Stevens, D. (2004) *Romanticism* (Cambridge: Cambridge University Press).

Stockman, N. (2000) *Understanding Chinese Society* (Cambridge: Cambridge University Press).

Stokes, M. (1996) 'Strong As a Turk: Power, Performance and Representation in Turkish Wrestling', in J. MacClancey (ed.) *Sport, Identity and Ethnicity* (Oxford: Berg).

Strange, S. (2004) *Stoicism: Traditions and Transformations* (Cambridge: Cambridge University Press).

Stroll, A. (2001) *Twentieth-century Analytical Philosophy* (New York: Columbia University Press).

Strutt, J. (1969[1801]) *The Sports and Pastimes of the People of England* (Bath: Firecrest).

Suetonius (2003) *The Twelve Caesars*, translated by R. Graves (Harmondsworth: Penguin).

Sugimoto, Y. (2009) *The Cambridge Companion to Modern Japanese Culture* (Cambridge: Cambridge University Press).

Sugimoto, Y. (2010) *An Introduction to Japanese Society* (Cambridge: Cambridge University Press).

Suits, B. (1977) 'Words on Play', *Journal of the Philosophy of Sport*, 4, 117–31.

Suits, B. (1988) 'On McBride on the Definition of Games', in W. Morgan and K. Meier (eds) *Philosophic Inquiry in Sport* (Champaign: Human Kinetics).

Suits, B. (2005) *The Grasshopper: Games, Life and Utopia* (Orchard Park: Broadview Press).

Sumption, J. (1999) *The Albigensian Crusade* (London: Faber and Faber).

Tang, C. (2008) *The Geographic Imagination of Modernity: Geography, Literature, and Philosophy in German Romanticism* (Palo Alto: Stanford University Press).

Thomas, K. (2003) *Religion and the Decline of Magic* (Harmondsworth: Penguin).

Thompson, P. (1983) 'The Structure of Evolutionary Theory: A Semantic Approach', *Studies in the History and Philosophy of Science*, 14, 215–29.

Thorndike, L. (1943) 'Renaissance or Prenaissance?', *Journal of the History of Ideas*, 4, 65–74.

Thorsteinsson, R. (2010) *Roman Christianity and Roman Stoicism* (Oxford: Oxford University Press).

Tilley, H. and Gordon, R. (2007) *Ordering Africa: Anthropology, European Imperialism and the Politics of Knowledge* (Manchester: Manchester University Press).

Toles-Patkin, T. (1986) 'Rational Co-ordination in the Dungeon', *Journal of Popular Culture*, 20, 1–14.

Tolkien, J.R.R. (1937) *The Hobbit* (London: Allen and Unwin).

Tolkien, J.R.R. (1954/55) *The Lord of the Rings* (London: Allen and Unwin).

Tolkien, J.R.R. (1977) *The Silmarillion* (London: Allen and Unwin).

Tolkien, J.R.R. (1983) *The Book of Lost Tales, Vol. One* (London: Allen and Unwin).

Tolkien, J.R.R. (1983) *The Book of Lost Tales, Vol. Two* (London: Allen and Unwin).

Totman, C. (2004) *History of Japan* (Oxford: Blackwell).

Toulmin, S.E. (1957) 'Crucial Experiments: Priestley and Lavoisier', *Journal of the History of Ideas*, 18, 205–20.

Toynbee, A. (1973) *Constantine Porphyrogenitus and His World* (Oxford: Oxford University Press).

Tribby, J. (1992) 'Body/Building: Living the Museum Life in Early Modern Europe', *Rhetorica: A Journal of the History of Rhetoric*, 10, 139–63.

Trigger, B. (2006) *A History of Archaeological Thought* (Cambridge: Cambridge University Press).

Tu, W. (1996) *Confucian Traditions in East Asian Modernity* (Cambridge: Harvard University Press).

Tulloch, J. and Jenkins, H. (1995) *Science Fiction Audiences: Watching Doctor Who and Star Trek* (London: Routledge).

Turner, F. (1993) *Contesting Cultural Authority: Essays in Victorian Intellectual Life* (Cambridge: Cambridge University Press).

Turner, V. (1982) *The Forest of Symbols* (Ithaca, New York: Cornell University Press).

Van Der Veer, P. (1994) *Religious Nationalism: Hindus and Muslims in India* (Sacramento: University of California Press).

Veblen, T. (1970[1899]) *The Theory of the Leisure Class* (London: Unwin).

Veblen, T. (2005) *Conspicuous Consumption* (Harmondsworth: Penguin).

Von Eschenbach, W. (1980) *Parzival*, translated by A. Hatto (Harmondsworth: Penguin).
Von Helden, I. (2010) 'Scandinavian Metal Attack: The Power of Northern Europe in Extreme Metal', in R. Hill and K. Spracklen (eds) *Heavy Fundametalisms: Music, Metal and Politics* (Oxford: ID-Net Press).
Wagg, S. (2004) 'Muck or Nettles: Men, Masculinity and Myth in Yorkshire Cricket', *Sport in History*, 23, 68–93.
Wagg, S. (2007) 'To be an Englishman: Nation, Ethnicity and English Cricket in the Global Age', *Sport in Society*, 10, 11–32.
Wagg, S. (2010) 'They Can't Stop Us Laughing: Politics, Leisure and the Comedy Business', in P. Bramham and S. Wagg (eds) *The New Politics of Leisure and Pleasure* (Basingstoke: Palgrave Macmillan).
Wainwright, A. (1955) *A Pictorial Guide to the Lakeland Fells: Book One: The Eastern Fells* (Kentmere: Henry Marshall).
Wainwright, A. (1957) *A Pictorial Guide to the Lakeland Fells: Book Two: The Far Eastern Fells* (Kentmere: Henry Marshall).
Wainwright, A. (1958) *A Pictorial Guide to the Lakeland Fells: Book Three: The Central Fells* (Kentmere: Henry Marshall).
Wainwright, A. (1960) *A Pictorial Guide to the Lakeland Fells: Book Four: The Southern Fells* (Kentmere: Henry Marshall).
Wainwright, A. (1962) *A Pictorial Guide to the Lakeland Fells: Book Five: The Northern Fells* (Kentmere: Henry Marshall).
Wainwright, A. (1964) *A Pictorial Guide to the Lakeland Fells: Book Six: The North Western Fells* (Kendal: Westmorland Gazette).
Wainwright, A. (1966) *A Pictorial Guide to the Lakeland Fells: Book Seven: The Western Fells* (Kendal: Westmorland Gazette).
Wainwright, A. (1968) *Pennine Way Companion* (Kendal: Westmorland Gazette).
Wainwright, A. (1986) *A Pennine Journey: The Story of a Long Walk in 1938* (London: Michael Joseph).
Wallbank, F. (2010) *The Hellenistic World* (London: Fontana).
Walton, J. (2000) *The British Seaside: Holidays and Resorts in the Twentieth Century* (Manchester: Manchester University Press).
Wang, N. (1999) 'Rethinking Authenticity in Tourism Experience', *Annals of Tourism Research*, 26, 349–70.
Ward, R. (1992) 'Women in Roman Baths', *Harvard Theological Review*, 85, 125–47.
Waring, E. (1969) *Rugby League: The Great Ones* (London: Pelham).
Warren, J. (2006) *Facing Death: Epicurus and His Critics* (London: Clarendon).
Warren, J. (2009) *The Cambridge Companion to Epicureanism* (Cambridge: Cambridge University Press).
Watts, E. (2004) 'Justinian, Malalas, and the End of Athenian Philosophical Teaching in AD 529', *The Journal of Roman Studies*, 94, 168–92.
Webb, D. (2002) *Medieval European Pilgrimage* (Basingstoke: Palgrave Macmillan).
Webb, D. (2007) *Pilgrimage in Medieval England* (Winchester: Hambledon Continuum).
Weber, M. (1992[1922]) *Economy and Society* (Sacramento: University of California Press).
Weber, M. (2001[1930]) *The Protestant Ethic and the Spirit of Capitalism* (London: Routledge).
Webster, C. (1982) *From Paracelsus to Newton* (Cambridge: Cambridge University Press).

Weeks, A. (1997) *Paracelsus: Speculative Theory and the Crisis of the Early Reformation* (Albany: State of New York University Press).
Weldes, J. (1999) 'Going Cultural: Star Trek, State Action and Popular Culture', *Millennium – Journal of International Studies*, 28, 117–34.
Wells, H.G. (1936) *A Short History of the World* (Harmondsworth: Penguin).
Wells, H.G. (2005[1895]) *The Time Machine* (Harmondsworth: Penguin).
Westfall, R. (1977) *The Construction of Modern Science: Mechanisms and Mechanics* (Cambridge; Cambridge University Press).
White, M. (1998) *Isaac Newton: The Last Sorcerer* (London: Fourth Estate).
Whittle, A. (2003) *The Archaeology of People: Dimensions of Neolithic Life* (London: Routledge).
Wickham, C. (2010) *The Inheritance of Rome: A History of Europe from 400 to 1000* (Harmondsworth: Penguin).
Williams, G. (1994) *The Code War: English Football Under the Historical Spotlight* (Harefield: Yore).
Williams, J. (1998) 'Cricket and Changing Expressions of Englishness', *Scottish Centre Research Papers*, 3, 101–12.
Williams, J. (1999) *Cricket and England: A Cultural and Social History of the Interwar Years* (London: Frank Cass).
Wilson, A. and Ashplant, T. (1988) 'Whig History and Present-Centred History', *The Historical Journal*, 31, 1–16.
Winter, J. (2000) *Tobacco use by Native North Americans: Sacred Smoke and Silent Killer* (Norman: University of Oklahoma Press).
Wittgenstein, L. (1968) *Philosophical Investigations* (Oxford: Blackwell).
Wolmar, C. (2010) *Blood, Iron and Gold: How the Railways Transformed the World* (London: Atlantic).
Wood, I. (1993) *The Merovingian Kingdoms, 450–571* (London: Longman).
Woolgar, C. Serjeantson, D. and Waldron, T. (2009) *Food in Medieval England: Diet and Nutrition* (Oxford: Oxford University Press).
Yao, X. (2000) *An Introduction to Confucianism* (Cambridge: Cambridge University Press).
Yapp, M. (1992) 'Europe in the Turkish Mirror', *Past and Present*, 137, 134–55.
Yates, F. (1964) *Giordano Bruno and the Hermetic Tradition* (London: University of Chicago Press).
Yates, F. (1972) *The Roscicrucian Enlightenment* (London: Routledge).
Young, D. (2005) 'Mens Sana in Corpore Sano? Body and Mind in Ancient Greece', *International Journal of the History of Sport*, 22, 22–41.
Yugul, F. (2009) *Bathing in the Roman World* (Cambridge: Cambridge University Press).
Yu-Lan, F. (1997) *A Short History of Chinese Philosophy: A Systematic Account of Chinese Thought from Its Origins to Present Day* (London: Simon and Schuster).
Zagorin, P. (1999) 'History, the Referent, and Narrative: Reflections on Postmodernism Now', *History and Theory*, 38, 1–24.
Zammito, J. (2005) 'Ankersmit and Historical Representation', *History and Theory*, 44, 155–81.
Zeeman, N. (2006) *Piers Plowman and the Medieval Discourse of Desire* (Cambridge: Cambridge University Press).

Index

America, 141, 153–4
Ammianus Marcellinus, 58–60
archaeological theory, 47
Aristotle, 18, 28, 32
authenticity, 6, 156
Aztecs, 40–2

Bacon, Francis, 23
ball game (Aztec), 41
baseball, 110
bathing, 54
Bentham, Jeremy, 19
Black Death, 98
board games, 38
Bollywood, 115–16
Bronze Age, 39–40
Byzantine Empire, 1–3, 68–71, 76–80

Caesar, Julius, 48–9, 55
Caliphate, Islamic, 75–6
Carnival, 130
Castiglione, Baldassare, 128
chariot racing, 2, 78
China, 107–10, 112–15
chivalry, 76, 88–9
Christianity, 22–3, 58–9, 71, 150
Chrysanthemum festivals, 109–10
Cicero, Marcus Tullius, 48, 60
circus factions, 2–4, 68–71
coca, 42
coffee shops, 142
Confucianism, 108–9
contact sports, 57
Copernicus, Nicolaus, 7
cosplay, 111
Cultural Revolution, Chinese, 114
Culture, The, 183
curry, 103

Darwin, Charles, 39
Descartes, Rene, 23–4
Dewey, John, 166–7
Domitian, 61
drinking, 21, 43–4, 54, 144

Enlightenment, The, 139, 141–4
Epicureanism, 21
evolution, 35–6

falconry, 92
feasting, 85, 88
folk dance, 155
football
  early, 87–8, 155
  modern, 112
Franklin, Benjamin, 141–2
Franks, The, 85–6

gambling, 21, 59
games, 25–6, 78–9
gaming, 111–12
gardens, 125
George, Henry, 144–8
Gibbon, Edward, 162–3
gin, 144
gladiators, 55–7
globalization, 110, 148, 178–9
Grand Tour, The, 142–3

Habermas, Jurgen, 5, 43, 52, 66, 86–8, 138, 142–4, 164–5
hajj, 73, 100
Hellenes, 51–2
Hermeticism, 124–5
Hobbes, Thomas, 24
Holy Days, 87
Holy Grail, 93–6
Huxley, Aldous, 184

iconoclasm, 71
Incas, 42
India, 103–4, 115–19
Islam, 71–6, 90, 91

Japan, 104–7, 110–12
Jordan, 38
Judge Dredd, 184–5
Justinian, 1–3
Juvenal, 4, 58

Kant, Immanuel, 24–5, 139

Languedoc, 96–7
Lascaux, 34, 37
leisure society thesis, 176
Levellers, The, 137
liquid leisure, 8, 177
London, 132–3

manners, 127–8
Marcus Aurelius, 21, 64
martial arts, 112
Marx, Karl, 148, 167
Mill, John Stuart, 27, 163–4
Milton, John, 162
modernity, 157–61
Mohammed, 73
music, 46, 61, 96–7, 104

nationalism, 154
Native Americans, 43
Neolithic Age, 38–9
Nero, 61, 64
New England, 135–6
Nietzsche, Friedrich, 25

Olympic Games, 52–3, 61, 149
orientalism, 103
Orwell, George, 184
otium, 59
Ottomans, 80–3, 154
Outremer, 91

Pausanias, 65–6
Petrarch, 126
Piers Plowman, 98
pilgrimage, 98–100
Plato, 20, 32, 53
play, 30
Popper, Karl, 18
printing, 129
Psellus, Michael, 77–80
puritanism, 135–8

Reformation, The, 122
Restoration England, 138
role-playing games, 192
Roman Empire, The, 54–5
Roman Republic, The, 53, 54
Rousseau, Jean-Jacques, 24, 44
rugby league, 151–3, 177–9

sagas, Icelandic, 89–90
Seneca, 56
sex, 21, 54, 60, 61
Shakespeare, William, 131–4
social justice theory, 27
Spain, 90
sport, definition of, 28–30
Star Trek, 185–8
stoicism, 21
Suits, Bernard, 30–1

tea, 103, 106–7
theatres, 132–4
Tiberius, 60–1
Tolkien, J.R.R., 188–91
troubadours, 93–7

Veblen, Thorstein, 168–9, 175
Venice, 128–31

walking, 169–72
Weber, Max, 168
Wells, H.G., 184
Whig history, 9–12
whisky, 119
Wittgenstein, Ludwig, 25–6, 31
world music, 172–4
Wuthering Heights, 119

York, 90